Fundamentals of Bayesian Epistemology 1

Fundamentals of Bayesian Epistemology 1

Introducing Credences

MICHAEL G. TITELBAUM

OXFORD
UNIVERSITY PRESS

OXFORD
UNIVERSITY PRESS

Great Clarendon Street, Oxford, OX2 6DP,
United Kingdom

Oxford University Press is a department of the University of Oxford.
It furthers the University's objective of excellence in research, scholarship,
and education by publishing worldwide. Oxford is a registered trade mark of
Oxford University Press in the UK and in certain other countries

Published in the United States of America by Oxford University Press
198 Madison Avenue, New York, NY 10016, United States of America

British Library Cataloguing in Publication Data
Data available

Library of Congress Control Number: 2021949533

ISBN 978-0-19-870760-8 (hbk.)
ISBN 978-0-19-870761-5 (pbk.)

DOI: 10.1093/oso/9780198707608.001.0001

Printed and bound by
CPI Group (UK) Ltd, Croydon, CR0 4YY

to Colleen
without whom nothing is possible

Contents

VOLUME 1

I. OUR SUBJECT

II. THE BAYESIAN FORMALISM

VOLUME 2

III. APPLICATIONS

IV. ARGUMENTS FOR BAYESIANISM

V. CHALLENGES AND OBJECTIONS

The most important questions of life . . . are indeed, for the most part, only problems in probability. One may even say, strictly speaking, that almost all our knowledge is only probable; and in the small number of things that we are able to know with certainty, in the mathematical sciences themselves, the principal means of arriving at the truth—induction and analogy—are based on probabilities, so that the whole system of human knowledge is tied up with the theory set out in this essay.

Pierre-Simon Laplace, *Philosophical Essay on Probabilities* (1814/1995)

Quick Reference

Non-Negativity: For any P in $\mathcal{L}$, $cr(P) \geq 0$.

Normality: For any tautology T in $\mathcal{L}$, $cr(T) = 1$.

Finite Additivity: For any mutually exclusive P and Q in $\mathcal{L}$, $cr(P \vee Q) = cr(P) + cr(Q)$.

Ratio Formula: For any P and Q in $\mathcal{L}$, if $cr(Q) > 0$ then $cr(P \mid Q) = \frac{cr(P \& Q)}{cr(Q)}$.

Conditionalization: For any time t_i and later time t_j, if E in $\mathcal{L}$ represents everything the agent learns between t_i and t_j, and $cr_i(E) > 0$, then for any H in $\mathcal{L}$, $cr_j(H) = cr_i(H \mid E)$.

CONSEQUENCES OF THESE RULES

Negation: For any P in $\mathcal{L}$, $cr(\sim P) = 1 - cr(P)$.

Maximality: For any P in $\mathcal{L}$, $cr(P) \leq 1$.

Contradiction: For any contradiction F in $\mathcal{L}$, $cr(F) = 0$.

Entailment: For any P and Q in $\mathcal{L}$, if $P \vDash Q$ then $cr(P) \leq cr(Q)$.

Equivalence: For any P and Q in $\mathcal{L}$, if $P =\!\vDash Q$ then $cr(P) = cr(Q)$.

General Additivity: For any P and Q in $\mathcal{L}$, $cr(P \vee Q) = cr(P) + cr(Q) - cr(P \& Q)$.

Finite Additivity (Extended): For any finite set of mutually exclusive propositions $\{P_1, P_2, \ldots, P_n\}$, $cr(P_1 \vee P_2 \vee \ldots \vee P_n) = cr(P_1) + cr(P_2) + \ldots + cr(P_n)$.

Decomposition: For any P and Q in $\mathcal{L}$, $cr(P) = cr(P \& Q) + cr(P \& \sim Q)$.

Partition: For any finite partition of propositions in $\mathcal{L}$, the sum of their unconditional cr-values is 1.

Law of Total Probability: For any proposition P and finite partition $\{Q_1, Q_2, \ldots, Q_n\}$ in $\mathcal{L}$,

$$cr(P) = cr(P \mid Q_1) \cdot cr(Q_1) + cr(P \mid Q_2) \cdot cr(Q_2) + \ldots + cr(P \mid Q_n) \cdot cr(Q_n).$$

Bayes's Theorem: For any H and E in $\mathcal{L}$,

$$cr(H \mid E) = \frac{cr(E \mid H) \cdot cr(H)}{cr(E)}.$$

Multiplication: P and Q with nonextreme cr-values are independent relative to cr if and only if $cr(P \& Q) = cr(P) \cdot cr(Q)$.

Preface

When I played junior varsity basketball in high school, our coach was always yelling at us about "the fundamentals". What my teammates and I derisively called "the fundies"—shooting, passing, dribbling—wouldn't win you a game by themselves. But without them, you'd have no chance of winning.

This book aims to teach you the fundamentals of Bayesian epistemology. It won't necessarily put you in a position to produce cutting-edge research in the area. But I hope that after reading this book you'll be able to pick up an article that uses the tools of Bayesian epistemology and understand what it says.

The word "Bayesian" is now familiar to academics across such diverse fields as statistics, economics, psychology, cognitive science, artificial intelligence, and even "legal probabilism" in the law. Bayesianism contains a few core notions across all of these fields: priors and posteriors, the probability calculus, updating by conditionalization. This book will introduce all of those core ideas.

Still, this book was written by a philosopher, and a philosopher with particular interests and areas of expertise. While I will make some allusions to those other fields, and will certainly draw connections to disparate parts of philosophy, I will focus on Bayesian *epistemology*—roughly, the idea that an agent has numerical degrees of belief subject to normative constraints based on probability mathematics. And I will examine Bayesian epistemology from a philosopher's point of view, asking the kinds of questions about it a philosopher would be interested in, and providing arguments in an analytic philosophical style.

For that reason, it will probably be easier to read this book if you have some background in philosophy (though that's not *strictly* necessary). As far as the mathematics goes, only a solid understanding of high-school level algebra will be assumed (except in Section 5.4). Chapter 2 quickly reviews the portions of elementary deductive logic—most of sentential logic, plus a bit of predicate logic—that will be needed later. If you've never had any exposure to formal logic, you might want to peruse an introductory text in that area before proceeding.

0.1 What's in this book

After this preface, the book has five parts. The first two parts appear in Volume 1, while the final three are in Volume 2. The first part, consisting of Chapter 1, defines a Bayesian epistemology as any theory adhering to two principles:

1. Agents have doxastic attitudes that can usefully be represented by assigning real numbers to claims.
2. Rational requirements on those doxastic attitudes can be represented by mathematical constraints on the real-number assignments closely related to the probability calculus.

Chapter 1 explains why it's helpful to include numerical degrees of belief (or "credences") in one's epistemology, instead of working exclusively with binary beliefs and/or confidence comparisons. This justifies the first principle of Bayesian epistemology.

The second part of the book then fleshes out the second principle, by laying out the normative rules of the Bayesian formalism. Chapter 2 covers the probability axioms, which represent rational constraints on an agent's unconditional degrees of belief at a given time. Chapter 3 examines conditional credences, relating them to unconditional credences through the Ratio Formula. Chapter 4 concerns Conditionalization, the traditional Bayesian norm for updating degrees of belief over time. Chapter 5 is a bit of a grab bag, introducing a variety of additional norms various Bayesians have proposed, none of which has achieved quite the canonical status of the rules covered in the previous chapters.

Part III investigates how the Bayesian formalism is *used*, covering its two most historically important applications. Chapter 6 applies Bayesianism to confirmation theory, the theory of how evidence justifies or supports hypotheses. Chapter 7 provides a brief introduction to decision theory (other texts go into much greater depth on this subject).

Once we've proposed a number of rational constraints on degrees of belief, and seen how they are used, Part IV asks how we might *argue* that these contraints are indeed normatively binding. A wide variety of arguments have been offered for the various constraints over the years; this part discusses three major argument-types. In Chapter 8, we see how any agent meeting certain plausible decision-theoretic constraints is representable as satisfying Bayesian norms. Chapter 9 looks at Dutch Books, which guarantee a sure loss for any agent violating the Bayesian rules. Chapter 10 covers more recent accuracy arguments, showing how Bayesian norms help an agent make her credences as accurate as possible.

Part V treats various challenges and objections to Bayesian epistemology. Chapter 11 explores how our Bayesian formalism might be extended to accommodate memory loss, and uncertainty about one's identity or spatio-temporal location. Chapter 12 takes up two interrelated objections to Bayesian norms: the "Problem of Old Evidence" that Bayesianism doesn't allow agents to learn now from information obtained in the past; and the "Problem of Logical Omniscience" that Bayesianism demands unrealistic logical abilities. Chapter 13 compares Bayesian epistemology to the statistical paradigms of frequentism and likelihoodism, focusing on the "Problem of the Priors" that is supposed to put Bayesianism at a serious disadvantage to those rivals. Finally, Chapter 14 considers three formalisms for representing intermediate levels of confidence—confidence comparisons, ranged credences, and Dempster-Shafer functions—that are related to the Bayesian template but try to improve it in various ways.

0.2 How to read—and teach—this book

Each chapter has the same structure. There's a brief introduction to the content of the chapter, then the main substance. In the main text, I **bold** a technical term when it is defined. I have opted for chapter endnotes rather than footnotes to highlight that the main text can be read through without attention to the notes. The notes contain references, technical extras for experts, random asides, and the occasional bad joke. After the main text, each chapter has a set of exercises, which are intended to review material from the chapter, challenge you to extend that material, and sometimes set up points for later chapters. Then there are recommendations for further reading, typically divided up into introductions and overviews of various topics from the chapter, classic texts from the Bayesian literature on those topics, and places where the discussion continues.

The frontmatter of each volume contains a Quick Reference page of Bayesian rules to which I'll be referring often. The backmatter contains three further resources: a glossary of terms, in which each term receives both a definition and a reference to the page on which it's initially defined; an index of names; and a bibliography.

I have successfully taught this book a number of times to upper-level undergraduates and early-year PhD students. I can teach almost the entire book in a semester. The pace is roughly one chapter per week, with extra time devoted to Chapter 5 (by far the most overstuffed chapter in the book), and a bit less time devoted to Chapter 8.

If you have to be selective in your reading—or your teaching—of this book, you should know that Chapters 2 through 4 provide formal material on which the entire rest of the book is based. Anyone who hasn't worked with the probability calculus before, or Bayesianism in general, should absolutely complete those chapters. (If you're in a serious hurry and want only the barest bones, the final section in each of those chapters—Sections 2.4, 3.3, and 4.3— may be skipped.) After that, the content becomes much more modular; each section of Chapter 5, for instance, may be read independently of the others. Some knowledge of decision theory (obtained via Chapter 7 or some other source) is necessary for Chapter 8, and probably helpful for Chapters 9 and 10. Some study of confirmation in Chapter 6 may also help with Chapter 13. But beyond those connections, each of the chapters after Part II was designed to depend only on the formal material from Part II; the farther you go in the book, the easier it should be just to skip to the chapters or even the sections of most interest.

It's also worth noting that probabilities move in counterintuitive ways. Intuitions built up from long experience working with full beliefs and deductive reasoning don't always transfer to degrees of belief and inductive contexts. People who work with probabilities for a living gradually build up new intuitions. Since I can't beam those directly into your brain, I've at least tried to point out situations in which relying on your old intuitions may be especially hazardous. Sections whose titles begin "Probabilities are weird!" focus especially on such situations. I also occasionally offer "Warning" boxes to highlight the most common confusions about Bayesianism I've encountered.

But the best way to build up new intuitions is to work through many probability exercises. Taking apart the guts of the Bayesian machinery and reassembling them to solve new problems is the best way to internalize how they work. So I recommend that anyone reading through this book—as a student or for their own interest—at least attempt some of the chapter exercises. Instead of boring, repetitive review of the content, I've tried to make the exercises intriguing, puzzling, and sometimes downright challenging. The level of challenge varies a great deal among the exercises, so I've used a system of one- to three-chili peppers to indicate degrees of difficulty. I've marked with a feather the exercises that call for essay-type responses; when I assign those to my classes I usually expect around a page apiece in response. Solutions to non-feather exercises are available to instructors by emailing the author. Still, I would *highly* recommend that instructors try solving exercises themselves before assigning them to students—especially the exercises with three chili peppers![1]

Having now written an introductory text, I have much greater respect for the authors of every introduction I've read before. The most difficult part was choosing what to leave in, and what to leave out. I quickly realized that I couldn't cover every move in every debate about Bayesianism. So my discussions aim to provide an entry point for readers into the critical conversation, rather than an all-encompassing summary of it. The entry point I provide has inevitably been shaped by my own perspective on the relevant issues. There are many ways to construct a Bayesian formalism; I have labored to present one here that is typical of the way most practitioners proceed.[2] Nevertheless, when it comes to interpreting and critiquing that formalism, my own concerns and ideas definitely come to the fore.

0.3 Acknowledgments

Below I've acknowledged everyone I can remember who provided advice on this book, answered my questions about the material, read a portion of it, taught it to their class, gave me notes on the text, or contributed in any other way. My apologies to anyone who should have appeared on this list but fell victim to my haphazard note-keeping process. Before providing the list, I want to single out a few resources and people for their special contributions. In writing the book I have relied on previous texts on aspects of Bayesianism, including Earman (1992), Skyrms (2000), Hacking (2001), Howson and Urbach (2006), Weisberg (2009), Bradley (2015), Talbott (2016), and unpublished lecture notes by Michael Strevens and Brian Weatherson. Galavotti (2005) is a phenomenal account of the history of Bayesian thought that I can't recommend highly enough. I am grateful to my editor, Peter Momtchiloff, who has been incredibly patient through this book's long gestation process; and to his Oxford University Press team and the referees and advisors whom he consulted along the way. Shimin Zhao read the final manuscript for me, wrote solutions to all the exercises, and improved the ultimate product in myriad ways. Shimin also checked the proofs with me. David Makinson encountered this book in a seminar he was attending, and without provocation conferred the great benefit of page-by-page comments on the entire manuscript as it existed at that time. Most importantly, Branden Fitelson introduced me to all of this material in the first place, shaped my thinking about it in ways I can't begin to recognize, and has been a dear friend and mentor ever since.

And now the list. My thanks to David Alexander, Dallas Amico, Yuval Avnur, Michael Barkasi, Joseph Barnes, Zach Barnett, Marty Barrett, Elizabeth Bell, John Bengson, JC. Bjerring, David Black, Darren Bradley, Seamus Bradley, R.A. Briggs, Stuart Brock, Lara Buchak, David Builes, Michael Caie, Catrin Campbell-Moore, Fabrizio Cariani, Jennifer Carr, Lisa Cassell, Clinton Castro, Jake Chandler, Isaac Choi, David Christensen, Hayley Clatterbuck, Nevin Climenhaga, Stew Cohen, Mark Colyvan, Juan Comesaña, Vincenzo Crupi, Martin Curd, Lars Dänzer, Glauber De Bona, Finnur Dellsén, Nick DiBella, Josh DiPaolo, Sinan Dogramaci, Kevin Dorst, Billy Dunaway, Kenny Easwaran, Philip Ebert, Andy Egan, Adam Elga, Jordan Ellenberg, Nic Fillion, Malcolm Forster, Melissa Fusco, Dmitri Gallow, Greg Gandenberger, Michał Godziszewski, Simon Goldstein, Daniel Greco, Max Griffin, Alba Guijarro, Alan Hájek, Jacqueline Harding, Casey Hart, Stephan Hartmann, Dan Hausman, Brian Hedden, Sophie Horowitz, Franz Huber, Liz Jackson, Pavel Janda, Jim Joyce, Mark Kaplan, Andrew Kernohan, Patrick Klug, Jason Konek, Matt Kopec, Johannes Korbmacher, Stephan Krämer, Jon Kvanvig, Tamar Lando, Bill Lawson, Hannes Leitgeb, Ben Lennertz, Ben Levinstein, Hanti Lin, Clayton Littlejohn, Tracy Lupher, Aidan Lyon, Amanda MacAskill, John MacFarlane, John Mackay, Anna Mahtani, Barry Maguire, Anna-Sara Malmgren, Conor Mayo-Wilson, David McCarthy, Tim McGrew, Brian McLoone, Chris Meacham, Silvia Milano, Peter Milne, Andrew Moon, Sarah Moss, Corey Mulvihill, André Neiva, Dilip Ninan, Ittay Nissan, Shannon Nolen, Samir Okasha, Jeff Paris, Sarah Paul, Carlotta Pavese, Trevor Pearce, Richard Pettigrew, Ted Poston, Vishal Prasad, Thomas Raleigh, Rosa Runhardt, Joe Salerno, Richard Samuels, Paolo Santorio, Joshua Schechter, Miriam Schoenfield, Jonah Schupbach, Teddy Seidenfeld, Glenn Shafer, Larry Shapiro, Alan Sidelle, Paul Silva, Rory Smead, Elliott Sober, Julia Staffel, Orri Stefánsson, Reuben Stern, Brian Talbot, David Thorstad, Brett Topey, Aron Vallinder, Brandt van der Gaast, Steven van Enk, Olav Vassend, Joel Velasco, Susan Vineberg, Justin Vlastis, Jonathan Vogel, Peter Vranas, Christian Wallmann, Petra Walta, Naftali Weinberger, Paul Weirich, Jonathan Weisberg, Roger White, Robbie Williams, Seth Yalcin; my students in Philosophy 504 and Philosophy 903 at the University of Wisconsin-Madison, and a Foundations seminar at the Australian National University; anyone I haven't yet listed who's ever attended the Formal Epistemology Workshop; my philosophy colleagues at the University of Wisconsin-Madison; and funding from the Australian National University, the Marc Sanders Foundation, and the following organizations affiliated with the University of Wisconsin-Madison: the Wisconsin Alumni Research Foundation, the Vilas Trust, the Office of the

Vice Chancellor for Research and Graduate Education, and the Institute for Research in the Humanities.

As the length of this list should indicate, philosophy is a community. It has been much critiqued—in part because it's our job to critique things, and in part because there are genuine, significant problems within that community. But it is also incredibly valuable, and that value should not be overlooked. Every piece of philosophy I have ever written has been dramatically improved by other members of the philosophy community: friends, colleagues, correspondents, audience members at conferences, etc. The formal epistemology community in particular has helped me at every step of this project, and is composed of a remarkably generous, intelligent, well-informed group of scholars.[3] Still, as this book will make apparent, there are many open questions right at the surface of our subject. If you finish the book and find yourself interested in pursuing any of them further, we invite you to join us.

Finally, I should add that while I am indebted to those listed here for their invaluable help, I am sure there remain errors in the main text, for which I assign sole responsibility to David Makinson.

Notes

1. Thanks to Joel Velasco for excellent advice on what teachers will want to know about this book, and to Maria Debowsky for the peppers icon and attendant LaTex code.
2. If you'd like to know how I personally prefer to build a formal Bayesian model, see Titelbaum (2013a).
3. At the beginning of his (2004), Richard Jeffrey recalls being introduced to Bayesianism by Carnap and Hempel. He writes that they were "the sweetest guys in the world. It seems to go with the territory." I couldn't agree more.

PART I
OUR SUBJECT

1

Beliefs and Degrees of Belief

Most of epistemology concerns propositional attitudes. A **propositional attitude** is an attitude an agent adopts toward a proposition, or toward a set of propositions. While much philosophical ink has been spilled over the nature of propositions, we will assume only that a **proposition** is an abstract entity expressible by a declarative sentence and capable of being true or false. (*True* and *false* are **truth-values**, so we say that a proposition is capable of "having a truth-value".) For example, the sentence "Nuclear fusion is a viable energy source" expresses a proposition. If I believe that fusion is viable, this belief is a propositional attitude—it is an attitude I take toward the proposition that fusion is viable.

Humans adopt a variety of attitudes toward propositions. I might *hope* that fusion is a viable energy source, *desire* that fusion be viable, *wonder whether* fusion is viable, *fear* that fusion is viable, or *intend to make it the case* that fusion is a viable energy source. While some propositional attitudes involve plans to change the world, others attempt to represent what the world is already like.

Epistemology focuses on the latter kind of propositional attitude—representational attitudes. Examples of such attitudes include belief and knowledge. (Knowledge will not be a major focus of this book.)[1] Belief is in some sense a *purely* representational attitude: when we attribute a belief to an agent, we are simply trying to describe how she takes the world to be. A belief attribution does not indicate any emotional affect toward the proposition, level of justification in that proposition, etc. Yet belief is not the only purely representational attitude; an agent might be *certain* that a proposition is true, or *disbelieve* a particular proposition. Philosophers often discuss the class of **doxastic attitudes** ("belief-like" attitudes) into which belief, disbelief, and certainty fall. Bayesian epistemology focuses on a type of doxastic attitude known variously as degree of belief, degree of confidence, or **credence**.

Over the last few decades discussion of credences has become much more common in epistemology, as well as in other areas of philosophy (not to mention psychology, economics, and adjacent disciplines). This chapter tries to explain why credences are important to epistemology. I'll begin by contrasting degree of belief talk with other doxastic attitude attributions—especially

Fundamentals of Bayesian Epistemology 1: Introducing Credences. Michael G. Titelbaum, Oxford University Press.
© Michael G. Titelbaum 2022. DOI: 10.1093/oso/9780198707608.003.0001

attributions of "binary" belief that have historically been significant in epistemology. I'll then consider what working with degrees of belief adds to our account of an agent's doxastic life. Finally I'll introduce a basic characterization of Bayesian epistemology, and outline how we will explore that view in the chapters to come.

1.1 Binary beliefs

1.1.1 Classificatory, comparative, quantitative

In his (1950), Rudolf Carnap helpfully distinguishes classificatory, comparative, and quantitative concepts:

> *Classificatory concepts* are those which serve for the classification of things or cases into two or a few [kinds].... *Quantitative concepts*...are those which serve for characterizing things or events or certain of their features by the ascription of numerical values.... *Comparative concepts*...stand between the two other kinds.... [They] serve for the formulation of the result of a comparison in the form of a more-less-statement without the use of numerical values.   (p. 9)

In Carnap's famous example, describing the air in a room as *warm* or *cold* employs classificatory concepts. Characterizing one room as *warmer* than another uses a comparative concept. The *temperature* scale describes the heat of a room with a quantitative concept.

Both our everyday talk about doxastic attitudes and our philosophical theorizing about them use classificatory, comparative, and quantitative concepts. Classificatory concepts include *belief, disbelief, suspension of judgment,* and *certainty*. The doxastic attitudes picked out by these concepts are monadic; each is adopted toward a single proposition. Moreover, given any particular proposition, agent, and classificatory doxastic attitude, the agent either has that attitude toward the proposition or she doesn't. So classificatory doxastic attitudes are sometimes called "binary". (I'll alternate between "classificatory" and "binary" terminology in what follows.) A comparative attitude, on the other hand, is adopted toward an ordered pair of propositions. For example, I am *more confident* that fission is a viable energy source than I am that fusion is. A quantitative attitude assigns a numerical value to a single proposition; my physicist friend is *90% confident* that fusion is viable.

Until the last few decades, much of epistemology revolved around classificatory concepts. (Think of debates about the justification of *belief*, or about necessary and sufficient conditions for *knowledge*.) This wasn't an exclusive focus, but more a matter of emphasis. So-called "traditional" or "mainstream" epistemologists certainly employed comparative and quantitative terms.[2] Moreover, their classificatory attitude ascriptions were subtly shaded by various modifiers: a belief, for example, might be *reluctant*, *intransigent*, or *deeply held*. Nevertheless, Bayesian epistemologists place much more emphasis on quantitative attitudes such as credences.

This chapter examines reasons for such a shift: Why should epistemologists be so interested in credences? To aid our understanding, I'll introduce a philosophical character who has probably never existed in real life: the **Simple Binarist**. A Simple Binarist insists on describing agents' doxastic propositional attitudes exclusively in terms of belief, disbelief, and suspension of judgment. The Simple Binarist eschews all other doxastic attitude attributions, and even refuses to add shading modifiers like the ones above. I introduce the Simple Binarist not as a plausible rival to the Bayesian, but instead as an illustrative contrast. By highlighting doxastic phenomena for which the Simple Binarist has trouble accounting, I will illustrate the importance of quantitative attitude attributions.

Nowadays most everyone uses a mix of classificatory, comparative, and quantitative doxastic concepts to describe agents' doxastic lives. I hope to demonstrate the significance of quantitative concepts within that mix by imagining what would happen if our epistemology lacked them entirely. And I will suggest that epistemologists' growing understanding of the advantages of degree-valued doxastic concepts helps explain the preponderance of quantitative attitude ascriptions in epistemology today.

1.1.2 Shortcomings of binary belief

My physicist friend believes that nuclear fusion is a viable energy source. She also believes that her car will stop when she presses the brake pedal. She is willing to bet her life on the latter belief, and in fact does so multiple times daily during her commute. She is not willing to bet her life on the former belief. This difference in the decisions she's willing to make seems like it should be traceable to a difference between her doxastic attitudes toward the proposition that fusion is viable and the proposition that pressing her brake pedal will stop her car. Yet the Simple Binarist—who is willing to attribute only beliefs, disbeliefs, and suspensions—can make out no difference between my friend's doxastic attitudes toward those propositions. Once the Simple Binarist says my friend believes both propositions, he has said all he has to say.

Now suppose my physicist friend reads about some new research into nuclear energy. The research reveals new difficulties with tokamak design, which will make fusion power more challenging. After learning of this research, she still believes fusion is a viable energy source. Nevertheless, it seems this evidence should cause *some* change in her attitude toward the proposition that fusion is viable. Yet the Simple Binarist lacks the tools to ascribe any such change; my friend believed the proposition before, and she still believes it now.

What do these two examples show? The Simple Binarist doesn't say anything *false*—it's *true* that my friend believes the propositions in question at the relevant times. But the Simple Binarist's descriptive resources don't seem fine-grained enough to capture some *further* things we want to say about my friend's doxastic attitudes. Now maybe there's some complicated way the Simple Binarist could account for these examples within his classificatory scheme. Or maybe a complex binarist with more classificatory attitudes in his repertoire than the Simple Binarist could do the trick. But it's most natural to respond to these examples with confidence *comparisons*: my friend is more confident that her brakes will work than she is that fusion is viable; reading the new research makes her less confident in the viability of fusion than she was before. Comparative doxastic attitudes fine-grain our representations in a manner that feels appropriate to these examples.

We've now seen two examples in which the Simple Binarist has trouble *describing* an agent's doxastic attitudes. But in addition to descriptive adequacy, we often want to work with concepts that figure in plausible *norms*.[3] Historically, epistemologists were often driven to work with comparative and quantitative doxastic attitudes because of their difficulties in framing defensible rational norms for binary belief.

The normative constraints most commonly considered for binary belief are:

Belief Consistency: Rationality requires the set of propositions an agent believes to be logically consistent.

Belief Closure: If some subset of the propositions an agent believes entails a further proposition, rationality requires the agent to believe that further proposition as well.

Belief Consistency and Belief Closure are proposed as necessary conditions for an agent's belief set to be rational. They are also typically proposed as requirements of *theoretical* rather than *practical* rationality.

Practical rationality concerns connections between attitudes and actions. Our earlier contrast between my friend's fusion beliefs and her braking beliefs was a practical one; it concerned how those doxastic attitudes influenced her betting *behavior*. Our other problematic example for the Simple Binarist was a purely theoretical one, having to do with my friend's fusion beliefs as evidence-responsive representations of the world (and without considering those beliefs' consequences for her actions).

What kinds of constraints does practical rationality place on attitudes? In Chapter 7 we'll see that if an agent's *preferences* fail to satisfy certain axioms, this can lead to a disastrous course of actions known as a "money pump". We might argue on this basis that practical rationality requires agents' preferences to satisfy those axioms. Similarly, we'll see in Chapter 9 that if an agent's *credences* fail to satisfy the probability axioms, her betting behavior is susceptible to a troublesome "Dutch Book". This fact has been used to argue that practical rationality requires credences to satisfy the probability axioms.

Perhaps practical rationality provides all the rational constraints that there are.[4] The standard response to this proposal invokes Pascal's Wager. Pascal (1670/1910, Section III) argues that it is rational to believe the proposition that the Christian god exists. If that proposition is true, having believed it will yield vast benefits in the afterlife. If the proposition is false, whether one believed it or not won't have nearly as dramatic consequences. Assuming Pascal has assessed these consequences correctly, they seem to provide some sort of reason to maintain religious beliefs. Nevertheless, if an agent's evidence points much more strongly to atheism than to the existence of a deity, it feels like there's a sense of rationality in which religious belief would be a rational mistake. This is **theoretical rationality**, a standard that assesses representational attitudes in their capacity as representations—how well they do at depicting the world, being responsive to evidence, etc.—without considering how they influence action. Belief Consistency and Closure are usually offered as requirements of theoretical rationality. The idea is that a belief set fails as a responsible representation of the world if it contradicts itself or doesn't include its own logical consequences.[5]

The versions of Belief Consistency and Closure I've stated above are pretty implausible as genuine rational requirements. Belief Closure, for instance, requires an agent to believe any arbitrarily complex proposition entailed by what she already believes, even if she's never come close to entertaining that proposition. And since any set of beliefs has infinitely many logical consequences, Closure also requires rational agents to have infinitely many beliefs. Belief Consistency, meanwhile, forbids an agent from maintaining a logically inconsistent set of beliefs even if the inconsistency is so recondite that she is incapable of recognizing it. One might find these requirements far too demanding to be rational constraints.

It could be argued, though, that Belief Consistency and Closure seem to have these flaws only because I've presented them inaptly or uncharitably. Perhaps we could make a few tweaks to the principles that would leave their spirit intact while inoculating them against these flaws. In Chapter 12 we will consider such tweaks to a parallel set of Bayesian constraints with similar problems. In the meantime, though, there are counterexamples to Belief Consistency and Closure that require much more than a few tweaks to resolve.

Kyburg (1961) first described the **Lottery Paradox**:

> A fair lottery has sold one million tickets. Because of the poor odds, an agent who has purchased a ticket believes her ticket will not win. She also believes, of each other ticket purchased in the lottery, that *it* will not win. Nevertheless, she believes that at least one purchased ticket *will* win.[6]

The beliefs attributed to the agent in the story seem rational. Yet these beliefs are logically inconsistent—you cannot consistently believe that at least one ticket will win while believing of each ticket that it will lose. So if the agent's beliefs in the story are rationally permissible, we have a counterexample to Belief Consistency. Moreover, if we focus just on the agent's beliefs about the individual tickets, that set of beliefs entails that none of the tickets will win. Yet it seems irrational for the agent to believe that no ticket will win. So the Lottery also provides a counterexample to Belief Closure.

Some defenders of Belief Consistency and Closure have responded that, strictly speaking, it is irrational for the agent in the Lottery to believe her ticket will lose. (If you believe your ticket will lose, why buy it to begin with?)[7] If true, this resolves the problem. But it's difficult to resolve Makinson's (1965) **Preface Paradox** in a similar fashion:

> You write a long nonfiction book with many claims in its main text, each of which you believe. In the acknowledgments at the beginning of the book you write, "While I am indebted to those listed here for their invaluable help, I am sure there remain errors in the main text, for which I take sole responsibility."

Many authors write such statements in the prefaces to their books, and it's hard to deny that it's rational for them to do so. It's also very plausible that nonfiction authors believe the contents of what they write. Yet if the concession that there are mistakes is an assertion that there is at least one falsehood in the main text, then the belief asserted in the preface is logically inconsistent with belief in all of the claims in the text.[8]

The Lottery and Preface pose a different kind of problem from our earlier examples. The examples with my friend the physicist didn't show that descriptions in classificatory belief terms were false; they simply suggested that classificatory descriptions don't capture all the important aspects of doxastic life. The Lottery and Preface, however, are meant to demonstrate that Belief Consistency and Belief Closure—the most natural normative principles for binary belief—are actually false.

An extensive literature has grown up around the Lottery and Preface, attempting to resolve them in a number of ways. One might deny that the sets of beliefs described in the paradoxes are in fact rational. One might find a clever way to establish that those sets of beliefs don't violate Belief Consistency or Belief Closure. One might drop Belief Consistency and/or Belief Closure for alternative normative constraints on binary belief. All of these responses have been tried, and I couldn't hope to adjudicate their successes and failures here.

For our purposes, the crucial point is that while it remains controversial how to square norms for binary belief with the Lottery and Preface, norms for rational credence have no trouble with those examples at all. In Chapter 2 we'll see that Bayesian norms tell a natural, intuitive story about the rational credences to adopt in the Lottery and Preface situations. The ease with which Bayesianism handles cases that are paradoxical for binary belief norms has been seen as a strong advantage for credence-centered epistemology.

1.2 From binary to graded

1.2.1 Comparative confidence

The previous section articulated both descriptive and normative difficulties for restricting one's attention exclusively to classificatory doxastic attitude ascriptions (belief, disbelief, suspension of judgment, etc.). We imagined a Simple Binarist who works only with these kinds of attitudes, and posed both descriptive and normative problems for him. The first descriptive problem was that an agent may believe two propositions while nevertheless treating these propositions quite differently when it comes to action. The second descriptive problem was that new evidence may change an agent's doxastic attitudes toward a proposition despite her believing the proposition both before and after incorporating the evidence. We could address both of these shortcomings in a natural fashion by moving beyond strictly classificatory concepts to *comparisons* between an agent's levels of confidence in two different propositions, or between her levels of confidence in a single proposition at two different times.

So let's augment our resources a bit beyond what the Simple Binarist has to offer. We'll still allow ourselves to say that an agent believes, disbelieves, or suspends judgment in a proposition. But we'll also allow ourselves to describe an agent as *at least as confident* of one proposition as another, *more confident* in one proposition than another, or *equally confident* in the two. Some of these comparisons follow directly from classificatory claims. For instance, when I say that my friend believes nuclear fusion is a viable energy source, we typically infer that she is more confident in the proposition that fusion is viable than she is in the proposition that fusion is nonviable. But there are also comparisons which, while consistent with classificatory information, are not entailed by such information. My friend believes both that fusion is viable and that her brakes are functional. We *go beyond* this description when we add that she is more confident in the latter proposition than the former.

We can think of an agent's confidence comparisons as aspects of an overall ranking of propositions. For example, Figure 1.1 depicts my confidence ranking of a particular set of propositions.[9] Here D represents the proposition that the Democrats will win the next presidential election, and W represents the proposition that anthropogenic global warming has occurred. The arrows indicate *more confident than* relations: for instance, I am more confident that warming either has or hasn't occurred than I am that it has, but I am also more confident that warming has occurred than I am that it has not.

It's important that not every confidence ranking is complete—there may be some pairs of propositions for which the ranking says nothing about the agent's relative confidences. Don't be fooled by the fact that "not D" and "W" are at the same height in Figure 1.1. In that diagram only the *arrows* reflect features of the ranking; the ranking depicted remains silent about whether I am more confident in "not D" or "W". This reflects an important truth about my doxastic attitudes: while I'm more confident in warming than nonwarming and in a Democratic loss than a win, I may genuinely be incapable of making a confidence comparison across those two unrelated issues. In

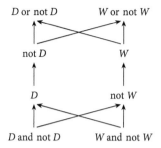

Figure 1.1 A confidence ranking

other words, I may view warming propositions and election propositions as *incommensurable*.

We now have the basic elements of a descriptive scheme for attributing comparative doxastic attitudes. How might we add a normative element to this scheme? A typical norm for confidence comparisons is:

Comparative Entailment: For any pair of propositions such that the first entails the second, rationality requires an agent to be at least as confident of the second as the first.

Comparative Entailment is intuitively plausible. For example, it would be irrational to be more confident in the proposition that Arsenal is the best soccer team in the Premier League than the proposition that Arsenal is a soccer team. Being the best soccer team in the Premier League *entails* that Arsenal is a soccer team![10]

Although it's a simple norm, Comparative Entailment has a number of substantive consequences. For instance, assuming we are working with a classical entailment relation on which any proposition entails a tautology and every tautology entails every other, Comparative Entailment requires a rational agent to be equally confident of every tautology and at least as confident of any tautology as she is of anything else. Comparative Entailment also requires a rational agent to be equally confident of every contradiction.

While Comparative Entailment (or something close to it)[11] has generally been endorsed by authors working on comparative confidence relations, there is great disagreement over which additional comparative norms are correct. Some alternatives are presented in Chapter 14, which works much more slowly and carefully through the technical details of comparative confidence rankings.

1.2.2 Bayesian epistemology

There is no single Bayesian epistemology; instead, there are many Bayesian *epistemologies*.[12] Every view I would call a Bayesian epistemology endorses the following two principles:

1. Agents have doxastic attitudes that can usefully be represented by assigning real numbers to claims.
2. Rational requirements on those doxastic attitudes can be represented by mathematical constraints on the real-number assignments closely related to the probability calculus.

The first of these principles is descriptive, while the second is normative—reflecting the fact that Bayesian epistemologies have both descriptive and normative commitments. Most of the rest of this chapter concerns the descriptive element; extensive coverage of Bayesian epistemology's normative content begins in Chapter 2.[13]

I've articulated these two principles vaguely to make them consistent with the wide variety of views (many of which we'll see later in this book) that call themselves Bayesian epistemologies. For instance, the first principle mentions "claims" because some Bayesian views assign real numbers to sentences or other entities in place of propositions. Still, the most common Bayesian descriptive approach—and the one we will stick to for most of this book—assigns numerical degrees of confidence to propositions.

In the previous section, we took the Simple Binarist's repertoire of belief, disbelief, and suspension descriptions and added comparative confidence relations. What more can we gain by moving to a full numerical representation of confidence? Confidence comparisons rank propositions, but cannot tell us how relatively large the gaps are between entries in the ranking. Without quantitative concepts we can say that an agent is more confident in one proposition than she is in another, but we cannot say *how much more* confident she is.

These matters of degree can be very important. Suppose you've been offered a job teaching at a university, but there's another university at which you'd much rather teach. The first university has given you two weeks to respond to their offer, and you know you won't have a hiring decision from the preferred school by then. Trying to decide whether to turn down the offer in hand, you contact a friend at the preferred university. She says you're one of only two candidates for their job, and she's more confident that you'll get the offer than the other candidate. At this point you want to ask *how much more* confident she is in your prospects than the other candidate's. A 51-49 split might not be enough for you to hang in!

Like our earlier brake pedal story, this is an example about the practical consequences of doxastic attitudes. It suggests that distinctions between doxastic attitudes affecting action cannot all be captured by a confidence ranking—important decisions may depend on the sizes of the gaps. Put another way, this example suggests that one needs more than just confidence comparisons to do decision theory (the subject of Chapter 7). In Chapter 6, we will use quantitative confidence measures to investigate a topic of great significance for theoretical rationality: degrees of confirmation. Numerical credence values are very important in determining whether a body of experimental evidence supports one scientific hypothesis more than it supports another.[14]

These are some of the advantages of numerically measuring degrees of belief. But credence ascriptions have disadvantages as well. For instance, numerical representations may provide more specific information than is actually present in the situation being represented. The Beatles were better than the Monkees, but there was no numerical *amount* by which they were better. Similarly, I might be more confident that the Democrats will lose the next election than I am that they will win without there being a fact of the matter about exactly *how much more* confident I am. Representing my attitudes by assigning precise credence values to the proposition that the Democrats will lose and the proposition that they will win attributes to me a confidence gap of a particular size—which may be an *over*-attribution in the actual case.

Numerical degree of belief representations also impose complete commensurability. It is possible to build a Bayesian representation that assigns credence values to some propositions but not others—representing the fact that an agent takes attitudes toward the former but not the latter.[15] But once our representation assigns a numerical credence to some particular proposition, that proposition immediately becomes comparable to every other proposition to which a credence is assigned. Suppose I am 60% confident that the Democrats will lose, 40% confident that they will win, and 80% confident that anthropogenic global warming has occurred. One can immediately rank all three of these propositions with respect to my confidence. Assigning numerical credences over a set of propositions creates a complete ranking, making it impossible to retain any incommensurabilities among the propositions involved. This is worrying if you think confidence incommensurability is a common and rational feature in real agents' doxastic lives.

Epistemologists sometimes complain that working with numerical credences is unrealistic, because agents "don't have numbers in their heads". This is a bit like refusing to measure gas samples with numerical temperature values because molecules don't fly around with numbers pinned to their backs.[16] The relevant question is whether agents' doxastic attitude sets have a structure that can be well represented by numbers, by a comparative ranking, by classificatory concepts, or by something else. This is the context in which it's appropriate to worry whether agents' confidence gaps have important size characteristics, or whether an agent's assigning doxastic attitudes to any two propositions should automatically make them confidence-commensurable. We will return to these issues a number of times in this book.

1.2.3   Relating beliefs and credences

I've said a lot about representing agents as having various doxastic attitudes. But presumably these attitudes aren't just things we can *represent* agents as having; presumably agents actually *have* at least some of the attitudes in question. The metaphysics of doxastic attitudes raises a plethora of questions. For instance: What *is* it—if anything—for an agent to genuinely possess a mental attitude beyond being usefully representable as having such? Or: If an agent can have both binary beliefs and degrees of belief in the same set of propositions, how are those different sorts of doxastic attitudes related? The latter question has generated a great deal of discussion, which I cannot hope to summarize here. Yet I do want to mention some of the general issues and best-known proposals.

Warning

There are two different philosophical conventions for employing the terms "belief" and "doxastic attitude", and it's important to disambiguate them before proceeding. In this book I will use "belief" as a synonym for "binary belief", one of the classificatory representational attitudes. "Doxastic attitude" will then be an umbrella term for propositional attitude types that are belief-like in the sense of being purely representational attitudes, including not only binary belief but also disbelief, certainty, doubt, suspension of belief, comparative confidence, numerical credence, and others. Yet there is a different convention on which "belief" is the umbrella term, and "doxastic attitude" means something like "variety of belief". On this approach, binary beliefs are sometimes called "full beliefs", and credences may be called "partial beliefs" or "graded beliefs". Adherents of this convention often say, "Belief comes in degrees." These last few locutions wouldn't make sense if "belief" meant exclusively binary belief. (A credence is not a partial or incomplete binary belief.) But they make more sense when "belief" is an umbrella term.

Going forward, I will refer to the quantitative representational attitudes that are our main topic as either "credences" or "degrees of belief".[17] I will also use "belief" and "doxastic attitude" according to the first of the two conventions just described, so for me "belief" will not be an umbrella term.

Now suppose some philosopher asserts a particular connection between (binary) beliefs and credences. That connection might do any of the following: (1) *define* attitudes of one kind in terms of the other; (2) *reduce* attitudes of one kind to attitudes of the other; (3) assert a *descriptively true* conditional (or biconditional) linking one kind of attitude to the other; (4) offer a *normative constraint* to the effect that any rational agent with an attitude of one kind will have a particular attitude of the other.

For example, the **Lockean thesis** connects believing a proposition with possessing a degree of confidence in that proposition surpassing some numerical threshold. Taking inspiration from John Locke (1689/1975, Bk. IV, Ch. 15–16), Richard Foley entertains the idea that:

> To say that you believe a proposition is just to say that you are sufficiently confident of its truth for your attitude to be one of belief. Then it is rational for you to believe a proposition just in case it is rational for you to have sufficiently high degree of confidence in it.   (1993, p. 140)

Foley presents the first sentence—identifying belief with sufficiently high degree of belief—as the Lockean thesis. The latter sentence is presented as following from the former. But notice that the latter sentence's normative claim could be secured by a weaker, purely normative Lockean thesis, asserting only that a *rational* agent believes a proposition just in case she is sufficiently confident of it.

On any reading of the Lockean thesis, there are going to be questions about exactly how high this threshold must be. One might suggest that the confidence threshold for belief is certainty (i.e., 100% confidence). But many of us believe propositions of which we are not certain, and this seems perfectly rational. Working down the confidence spectrum, it seems that in order to believe a proposition one should be more confident of it than not. But that leaves a lot of space to pin down the threshold between 50% and 100% confidence. Here it may help to suggest that the relevant threshold for belief is vague, or varies with context.

The Lockean thesis also causes problems when we try to layer traditional norms of rational belief and credence on top of it. If we set a credence threshold for belief lower than 100%, and adopt Bayesian probabilistic norms for credence, the Lockean thesis generates rational belief sets for the Lottery and Preface that violate Belief Consistency and Closure. We will see why when we give a probabilistic solution to the Lottery in Section 2.2.2.

The Lockean thesis works by identifying belief with a particular kind of credence. But we might try connecting these attitudes in the opposite direction—identifying credence with a particular kind of belief. For instance, we might say that I have a 60% credence that the Democrats will lose the next election just in case I believe the proposition that their probability of losing is 60%. The general strategy here is to align my credence in one proposition with belief in a second proposition about the *probability* of the first.

This connective strategy—whether meant definitionally, reductively, normatively, etc.—is now generally viewed as unlikely to succeed. For one thing, it requires thinking that whenever a (rational) agent has a degree of confidence, she also has a belief about probabilities. David Christensen (2004, Ch. 2) wonders about the content of these probability beliefs. In Chapter 5 we will explore various "interpretations of probability" that attempt to explain the meaning of "probability" claims. The details need not concern us here; what matters is that for each possible interpretation, it's implausible to think that whenever a (rational) agent has a degree of confidence she (also?) has a belief with that kind of probabilistic content. If "probability" talk is, for instance, always talk about frequency within a reference class, must I have beliefs about frequencies and reference classes in order to be pessimistic about the Democrats' prospects?

The idea that the numerical value of a credence occurs inside a proposition toward which the agent adopts some attitude also generates deeper problems. We will discuss some of them when we cover conditional credences in Chapter 3. Generally, contemporary Bayesians think of the numerical value of a credence not as part of the content toward which the agent adopts the attitude, but instead as an attribute of the attitude itself. I adopt a credence *of 60%* toward the proposition that the Democrats will lose; no proposition *containing* the value 60% is involved.[18]

This is a small sample of the positions and principles that have been proposed relating beliefs to degrees of belief. One might embrace some connecting principle I haven't mentioned here. Or one might deny the existence of attitudes in one category altogether. (Perhaps there are no beliefs. Perhaps there are no degrees of belief.) Yet I'd like to note that it is possible that both types of attitudes exist without there being any fully general, systematic connections between them.

Here's an anology:[19] Consider three different maps of the same square mile of earthly terrain. One is a topographic map; another is a satellite image; another shows streets marked with names. Each map represents different features of the underlying terrain. The features represented on each map are equally real. There are *some* connections between the information on one map and the information on another; a street that appears on the satellite photo will presumably appear on the streetmap as well. But there are no fully general, systematic connections that would allow you to derive *everything* about one map from any of the others. For instance, nothing on the topo or the streetmap provides the location of a tree picked up by the satellite.

Similarly, describing agents as possessing beliefs or as possessing degrees of belief might be equally valid representations of a complex underlying reality, with each description useful for different purposes. The features of an agent's cognitive state picked out by each representation might also be equally real. Yet there might nevertheless be no general, systematic connections between one representation and the other (even for a fully rational agent). Going forward, we will assume that it is at least sometimes philosophically useful to represent agents as having numerical degrees of belief. We will not assume any systematic connection between credences and beliefs, and indeed we will only rarely mention the latter.

1.3 The rest of this book

Hopefully I have now given you some sense of what credences are, and of why one might incorporate them into one's epistemology. Our first task in Chapter 2 will be to develop a Bayesian formalism in which credences can be descriptively represented. After that, much of our focus will be on the norms Bayesians require of rational degrees of belief.

There is a great deal of disagreement among Bayesians about exactly what these norms should be. Nevertheless, we can identify five core normative Bayesian rules: Kolmogorov's three probability axioms for unconditional credence, the Ratio Formula for conditional credence, and Conditionalization for updating credences over time. These are not core rules in the sense that all Bayesian epistemologists agree with them. Some Bayesians accept all five rules and want to add more; some don't even accept these five. They are core in the sense that one needs to understand them in order to understand any further Bayesian position entertained.

This chapter completes Part I of this book. Part II is primarily concerned with the five core Bayesian rules. Chapter 2 covers Kolmogorov's axioms; Chapter 3 covers the Ratio Formula; and Chapter 4 covers Conditionalization. Chapter 5 then discusses a variety of norms Bayesians have proposed either to supplement or to replace the core five.

The presence of all these alternatives raises the question of why we should accept any of these rules as genuinely normative to begin with. To my mind, one can see the advantages of Bayesianism best by seeing its consequences for applications. For instance, I've already mentioned that Bayesian credence norms accommodate a natural story about doxastic attitudes in the Lottery Paradox. Part III of this book discusses the two historically most important applications of Bayesian epistemology: confirmation theory (Chapter 6) and decision theory (Chapter 7).

Along with their benefits in application, Bayesian normative rules have been directly defended with a variety of philosophical arguments. I discuss the three most popular arguments in Part IV, and explain why I find each ultimately unconvincing. Chapter 8 discusses Representation Theorem arguments; Chapter 9 Dutch Books; and Chapter 10 arguments based on the goal of having accurate credences.

Finally, a number of important challenges have been raised to Bayesian epistemology—both to its descriptive framework and to its normative rules. Many of these (though not nearly all) are covered in Part V. Chapter 11 discusses how we might extend Bayesian updating to incorporate memory loss and self-locating belief. Chapter 12 takes up the Problem of Old Evidence and the Problem of Logical Omniscience. Chapter 13 introduces the Problem of the Priors, then compares Bayesianism to the rival statistical paradigms of frequentism and likelihoodism. Finally, Chapter 14 considers alternative frameworks for modeling levels of confidence that employ the probability calculus but put it to different formal use.

1.4 Exercises

Problem 1.1. 🖋 What do *you* think the agent in the Lottery Paradox should believe? In particular, should she believe of each ticket in the lottery that that ticket will lose? Does it make a difference how many tickets there are in the lottery? Explain and defend your answers.

Problem 1.2. 🎵 Explain why (given a classical logical entailment relation) Comparative Entailment requires a rational agent to be equally confident of every contradiction.

Problem 1.3. 🎵 Assign numerical confidence values (between 0% and 100%, inclusive) to each of the propositions in Figure 1.1. These confidence values should be arranged so that if there's an arrow in Figure 1.1 from one proposition to another, then the first proposition has a lower confidence value than the second.

Problem 1.4. 🎵🎵 The arrows in Figure 1.1 represent "more confident in" relations between pairs of propositions. Comparative Entailment, on the other hand, concerns the "at least as confident in" relation. So suppose we reinterpreted Figure 1.1 so that the arrows represented "at least as confident in" relations. (For example, Figure 1.1 would now tell you that I'm at least as confident of a Democratic loss as a win.)
  (a) Explain why—even with this reinterpretation—the arrows in Figure 1.1 do not provide a ranking that satisfies Comparative Entailment.
  (b) Describe a bunch of arrows you could add to the (reinterpreted) diagram to create a ranking satisfying Comparative Entailment.

Problem 1.5. 🖋 Is it *ever* helpful to describe an agent's attitudes in terms of binary beliefs? Or could we get by just as well using only more fine-grained (comparative and quantitative) concepts? Explain and defend your answer.

1.5 Further reading

INTRODUCTIONS AND OVERVIEWS

Elizabeth G. Jackson (2020). The Relationship between Belief and Credence. *Philosophy Compass* 15, pp. 1–13

Systematic overview of the various philosophical views (both descriptive and normative) connecting beliefs and credences. Includes a nice discussion of the possible metaphysical relationships between the two types of attitude.

CLASSIC TEXTS

Henry E. Kyburg Jr (1970). Conjunctivitis. In: *Induction, Acceptance, and Rational Belief.* Ed. by M. Swain. Boston: Reidel, pp. 55–82
David C. Makinson (1965). The Paradox of the Preface. *Analysis* 25, 205–7

Classic discussions of the Lottery and Preface Paradoxes (respectively), by the authors who introduced these paradoxes to the philosophical literature.

EXTENDED DISCUSSION

Richard Foley (2009). Beliefs, Degrees of Belief, and the Lockean Thesis. In: *Degrees of Belief.* Ed. by Franz Huber and Christoph Schmidt-Petri. Vol. 342. Synthese Library. Springer, pp. 37–48
Ruth Weintraub (2001). The Lottery: A Paradox Regained and Resolved. *Synthese* 129, pp. 439–49
David Christensen (2004). *Putting Logic in its Place.* Oxford: Oxford University Press

Foley, Weintraub, and Christensen each discuss the relation of binary beliefs to graded, and the troubles for binary rationality norms generated by the Lottery and Preface Paradoxes. They end up leaning in different directions: Christensen stresses the centrality of credence to norms of theoretical rationality, while Foley and Weintraub emphasize the role of binary belief in a robust epistemology.

Notes

1. While Bayesian epistemology has historically focused on *doxastic* representational attitudes, some authors have recently applied Bayesian ideas to the study of knowledge. See, for instance, Moss (2018).
2. John Bengson, who has greatly helped me with this chapter, brought up the interesting historical example of how we might characterize David Hume's (1739–40/1978) theory of belief vivacity in classificatory/comparative/quantitative terms.
3. On some epistemologies, descriptive and normative projects cannot be prized apart, because various normative conditions are either definitional of or constitutive of what

it *is* to possess particular doxastic attitudes. See, for instance, Davidson (1984) and Kim (1988).

4. See, for example, Kornblith (1993). Kornblith has a response to the Pascalian argument I'm about to offer, but chasing down his line would take us too far afield.

5. Has Pascal conclusively demonstrated that practical rationality requires religious belief? I defined practical rationality as concerning an attitude's connection to action. One odd aspect of Pascal's Wager is that it seems to treat believing as a kind of action in itself. Many philosophers have wondered whether we have the kind of direct control over our beliefs to deliberately follow Pascal's advice.

   For our purposes, the crucial point is that the pressure to honor atheistic evidence doesn't seem immediately connected to action. This suggests a standard of theoretical rationality distinct from concerns of practical rationality.

6. William Folz pointed out to me that when many American English speakers hear the word "lottery", they think of something like a state lotto (scratch-off tickets, Keno, etc.), in which one's chances of winning are often unaffected by the number of other players. Decision theorists, statisticians, and philosophers of probability (not to mention Shirley Jackson) use the word "lottery" to cover a wider variety of chance events. The kind of lottery described in the Lottery Paradox (and at other points in this book) might more commonly be called a "raffle".

7. This is why I never play the lottery.

8. If you find the Preface Paradox somehow unrealistic or too distant from your life, consider that (1) you have a large number of beliefs (each of which, presumably, you believe); and (2) you may also believe (quite reasonably) that at least one of your beliefs is false. This combination is logically inconsistent.

9. One of the challenges of introducing technical material in ordinary English is that many of the words of ordinary English have also been assigned technical meanings by mathematicians or philosophers. If one looks, one can find formal definitions of what constitutes a "ranking"—as well as an entire formalism called "ranking theory". Nevertheless, I am going to use "ranking" in an informal sense throughout this book, to indicate the general idea of placing some things higher than others, some lower, and some on equal footing.

10. In an article dated January 2, 2014 on grantland.com, a number of authors made bold predictions for the forthcoming year. Amos Barshad wrote:

    And so, here goes, my two-part prediction:
    1. The Wu-Tang album will actually come out.
    2. It'll be incredible.
    I'm actually, illogically more sure of no. 2.

11. Comparative Entailment shares some of the intuitive flaws we pointed out earlier for Belief Closure: (1) as stated, Comparative Entailment requires an agent to compare infinitely many ordered pairs of propositions (including propositions the agent has never entertained); (2) Comparative Entailment places demands on agents who have not yet recognized that some particular proposition entails another. So it is tempting to tweak Comparative Entailment in ways similar to the tweaks we will propose in Chapter 12 for Belief Consistency, Belief Closure, and their Bayesian cousins.

12. I.J. Good famously argued in a letter to the editor of *The American Statistician* that there are at least 46,656 varieties of Bayesians (Good 1971).

13. These days philosophers sometimes talk about "Formal Epistemology". A formal epistemology is any epistemological theory that uses formal tools. Bayesian epistemology is just one example of a formal epistemology; other examples include AGM theory (Alchourrón, Gärdenfors, and Makinson 1985) and the aforementioned ranking theory (Spohn 2012).

14. There exists a quantitative strand of epistemology (e.g., Pollock 2001) focusing on numerical degrees of *justification* conferred by bodies of evidence. Since this topic is potentially related to degrees of confirmation, I will return to it in Section 6.4.1. For now, it suffices to note that even if an agent's evidence confers some numerically measurable amount of justification on a particular proposition or a particular belief for her, that degree of justification is conceptually distinct from her degree of belief in the proposition, and the norms we will study apply to the latter.

15. I'll describe some details of this construction in Chapter 12.

16. Joel Velasco reminded me that doctors often ask us to rate our pain on a scale of 1 to 10. May we respond only if we have numbers in our heads? In our nerves?

17. The word "credence" has been used as a mass noun in English for centuries, coming over from French and ultimately having Latin roots. It's unclear to me, though, when it started being used as a count noun the way Bayesians do—as when we say that agents have credences in particular claims. In a discussion on this subject with Branden Fitelson and Andrew Moon, Fitelson highlighted the following passage in Carnap that may supply an important bridge to the contemporary usage: "The concept of probability in the sense of the actual degree of belief is a psychological concept.... I shall use for this psychological concept the technical term 'degree of credence' or shortly 'credence'" (Carnap 1962b, p. 305).

18. If we shouldn't think of the number in a numerical credence as part of the content of the proposition toward which the attitude is adopted, how exactly *should* we think of it? I tend to think of the numerical value as a sort of property or adjustable parameter of a particular doxastic attitude-type: credence. An agent adopts a credence toward a specific proposition, and it's a fact about that credence that it has degree 60% (or whatever).

    For the contrasting view, and arguments in favor of putting the numerical value in the content of a proposition believed, see Lance (1995) and Holton (2014). Another option—which would take us too far afield to address in this book—is to read a credence as a belief in a more complex kind of content, one component of which is propositional and a *distinct* component of which is numerical. Moss (2018) adopts this approach.

19. Thanks to Elizabeth Bell for discussion. I also found the following passage from Lange (2019) evocative: "In his landmark book, *The Image of the City*, published in 1961, [Kevin] Lynch asked people to draw their city for a visitor, paying attention to their own everyday paths and major landmarks, without reference to geography. Of course, each person's map...was different—but that did not mean that one map was more accurate than another. Rather, each person was telling a different story through cartography."

PART II
THE BAYESIAN FORMALISM

There are five core normative rules of Bayesian epistemology: Kolmogorov's three probability axioms, the Ratio Formula, and updating by Conditionalization. That is not to say that these are the only normative rules Bayesians accept, or that all Bayesians accept all five of these. But one cannot understand any additional rules or replacement rules without understanding these five first.

Chapter 2 begins with some review of propositions and propositional logic. It then discusses unconditional credence, an agent's general degree of confidence that a particular proposition is true. The Kolmogorov axioms are introduced as rational constraints on unconditional credences, then their consequences are explored. Finally, I discuss how the resulting normative system goes beyond what one gets from simple non-numerical norms for comparative confidence.

Chapter 3 then introduces *conditional* credence—an agent's confidence that one proposition is true on the supposition that another proposition is. The Ratio Formula is a normative rule relating an agent's conditional credences to her unconditional credences. Chapter 3 applies the Ratio Formula to develop Bayesian notions of relevance and probabilistic independence. It then discusses relationships among conditional credences, causes, and conditional propositions.

The probability axioms and the Ratio Formula relate credences held by an agent at a given time to other credences held by that agent at the same time. Conditionalization relates an agent's credences at different times. After introducing Conditionalization, Chapter 4 discusses the roles that evidence and certainty play in that rule. It then explains how Conditionalization does the useful epistemological work of distinguishing an agent's *evidence* from the *epistemic standards* she brings to bear on that evidence.

Chapter 5 begins by discussing distinctions between "Subjective" and "Objective" Bayesians, and various interpretations of "probability" talk. It then covers a number of well-known Bayesian norms that go beyond the core five, including: the Principal Principle, the Reflection Principle, various other deference principles, the Principle of Indifference, Countable Additivity, and Jeffrey Conditionalization.

2

Probability Distributions

This chapter introduces Kolmogorov's probability axioms, the first three core normative rules of Bayesian epistemology. They represent constraints that an agent's unconditional credence distribution at a given time must satisfy in order to be rational.

The chapter begins with a quick overview of propositional and predicate logic. The goal is to remind readers of logical notation and terminology we will need later; if this material is new to you, you can learn it from any introductory logic text. Next I introduce the notion of a numerical distribution over a propositional language, the tool Bayesians use to represent an agent's degrees of belief. Then I present the probability axioms, which are mathematical constraints on such distributions.

Once the probability axioms are on the table, I point out some of their more intuitive consequences. The probability calculus is then used to analyze the Lottery Paradox scenario from Chapter 1, and Tversky and Kahneman's Conjunction Fallacy example.

Kolmogorov's axioms are the canonical way of *defining* what it is to be a probability distribution, and they are useful for doing probability proofs. Yet there are other, equivalent mathematical structures that Bayesians often use to illustrate points and solve problems. After presenting the axioms, this chapter describes how to work with probability distributions in three alternative forms: Venn diagrams, probability tables, and odds.

I end the chapter by explaining what I think are the most distinctive elements of probabilism, and how probability distributions go beyond what one obtains from a comparative confidence ranking.

2.1 Propositions and propositional logic

Following the discussion in Chapter 1, we will assume that degrees of belief are propositional attitudes—they are attitudes agents assign to propositions.[1] In any particular application we will be interested in the degrees of belief an agent assigns to the propositions in some language $\mathcal{L}$. $\mathcal{L}$ will contain a finite

Fundamentals of Bayesian Epistemology 1: Introducing Credences. Michael G. Titelbaum, Oxford University Press.
© Michael G. Titelbaum 2022. DOI: 10.1093/oso/9780198707608.003.0002

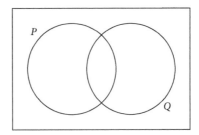

Figure 2.1 The space of possible worlds

number of **atomic propositions**, which we will usually represent with capital letters (P, Q, R, etc.).

The rest of the propositions in $\mathcal{L}$ are constructed in standard fashion from atomic propositions using five **propositional connectives**: $\sim$, &, $\vee$, $\supset$, and $\equiv$. A **negation** $\sim P$ is true just in case P is false. A **conjunction** $P \,\&\, Q$ is true just in case its **conjuncts** P and Q are both true. "$\vee$" represents inclusive "or"; a **disjunction** $P \vee Q$ is false just in case its **disjuncts** P and Q are both false. "$\supset$" represents the **material conditional**; $P \supset Q$ is false just in case its **antecedent** P is true and its **consequent** Q is false. A **material biconditional** $P \equiv Q$ is true just in case P and Q are both true or P and Q are both false.

Philosophers sometimes think about propositional connectives using sets of **possible worlds**. Possible worlds are somewhat like the alternate universes to which characters travel in science-fiction stories—events occur in a possible world, but they may be different events than occur in the **actual world** (the possible world in which *we* live). Possible worlds are maximally specified, such that for any event and any possible world that event either does or does not occur in that world. And the possible worlds are plentiful enough such that for any combination of events that *could* happen, there is a possible world in which that combination of events *does* happen.

We can associate with each proposition the set of possible worlds in which that proposition is true. Imagine that in the **Venn diagram** of Figure 2.1 (named after a logical technique developed by John Venn), the possible worlds are represented as points inside the rectangle. Proposition P might be true in some of those worlds, false in others. We can draw a circle around all the worlds in which P is true, label it P, and then associate proposition P with the set of all possible worlds in that circle (and similarly for proposition Q).

The propositional connectives can also be thought of in terms of possible worlds. $\sim P$ is associated with the set of all worlds lying outside the P-circle. $P \,\&\, Q$ is associated with the set of worlds in the overlap of the P-circle and the Q-circle. $P \vee Q$ is associated with the set of worlds lying in either the P-circle

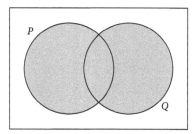

Figure 2.2 The set of worlds associated with $P \vee Q$

or the Q-circle. (The set of worlds associated with $P \vee Q$ has been shaded in Figure 2.2 for illustration.) $P \supset Q$ is associated with the set containing all the worlds except those that lie both inside the P-circle and outside the Q-circle. $P \equiv Q$ is associated with the set of worlds that are either in both the P-circle and the Q-circle or in neither one.[2]

Warning

I keep saying that a proposition can be "associated" with the set of possible worlds in which that proposition is true. It's tempting to think that the proposition just *is* that set of possible worlds, but we will avoid that temptation. Here's why: The way we've set things up, any two logically equivalent propositions (such as P and $\sim P \supset P$) are associated with the same set of possible worlds. So if propositions just *were* their associated sets of possible worlds, P and $\sim P \supset P$ would be the same proposition. Since we're taking credences to be assigned to propositions, that would mean that *of necessity* every agent assigns P and $\sim P \supset P$ the same credence. Eventually we're going to suggest that if an agent assigns P and $\sim P \supset P$ different credences, she's making a rational mistake. But we want our formalism to deem it a *rational requirement* that agents assign the same credence to logical equivalents, not a *necessary truth*. It's useful to think about propositions in terms of their associated sets of possible worlds, so we will continue to do so. But to keep logically equivalent propositions separate entities we will not say that a proposition just is a set of possible worlds.

Before we discuss relations among propositions, a word about notation. I said we will use capital letters to represent specific atomic propositions. We will also use capital letters as metavariables ranging over propositions. I might say,

"If P entails Q, then…". Clearly the atomic proposition P doesn't entail the atomic proposition Q. So what I'm saying in such a sentence is "Suppose we have one proposition (which we'll call 'P' for the time being) that entails another proposition (which we'll call 'Q'). Then…". At first it may be confusing sorting atomic proposition letters from metavariables, but context will hopefully make my usage clear. (Look out especially for phrases like: "For any propositions P and Q…".)[3]

2.1.1 Relations among propositions

Propositions P and Q are **equivalent** just in case they are associated with the same set of possible worlds—in each possible world, P is true just in case Q is. In that case I will write "$P \dashv\vDash Q$". P **entails** Q ("$P \vDash Q$") just in case there is no possible world in which P is true but Q is not. On a Venn diagram, P entails Q when the P-circle is entirely contained within the Q-circle. (Keep in mind that one way for the P-circle to be entirely contained in the Q-circle is for them to be the same circle! When P is equivalent to Q, P entails Q and Q entails P.) P **refutes** Q just in case $P \vDash \sim Q$. When P refutes Q, every world that makes P true makes Q false.[4]

For example, suppose I have rolled a six-sided die. The proposition that the die came up six entails the proposition that it came up even. The proposition that the die came up six refutes the proposition that it came up odd. The proposition that the die came up even is equivalent to the proposition that it did not come up odd—and each of those propositions entails the other.

P is a **tautology** just in case it is true in every possible world. In that case we write "$\vDash P$". I will sometimes use the symbol "T" to stand for a tautology. A **contradiction** is false in every possible world. I will sometimes use "F" to stand for a contradiction. A **contingent** proposition is neither a contradiction nor a tautology.

Finally, we have properties of proposition *sets* of arbitrary size. The propositions in a set are **consistent** if there is at least one possible world in which all of those propositions are true. The propositions in a set are **inconsistent** if no world makes them *all* true.

The propositions in a set are **mutually exclusive** if no possible world makes *more than one* of them true. Put another way, each proposition in the set refutes each of the others. (For any propositions P and Q in the set, $P \vDash \sim Q$.) The propositions in a set are jointly **exhaustive** if each possible world makes at

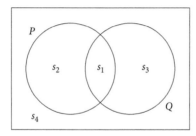

Figure 2.3 Four mutually exclusive, jointly exhaustive regions

least one of the propositions in the set true. In other words, the disjunction of all the propositions in the set is a tautology.

We will often work with proposition sets whose members are both mutually exclusive and jointly exhaustive. A mutually exclusive, jointly exhaustive set of propositions is called a **partition**. Intuitively, a partition is a way of dividing up the available possibilities. For example, in our die-rolling example the proposition that the die came up odd and the proposition that the die came up even together form a partition. When you have a partition, each possible world makes *exactly* one of the propositions in the partition true. On a Venn diagram, the regions representing the propositions in a partition combine to fill the entire rectangle without overlapping at any point.

2.1.2 State-descriptions

Suppose we are working with a language that has just two atomic propositions, P and Q. Looking back at Figure 2.1, we can see that these propositions divide the space of possible worlds into four mutually exclusive, jointly exhaustive regions. Figure 2.3 labels those regions s_1, s_2, s_3, and s_4. Each of the regions corresponds to one of the lines in the following truth-table:

	P	Q	state-description
s_1	T	T	$P \& Q$
s_2	T	F	$P \& {\sim}Q$
s_3	F	T	${\sim}P \& Q$
s_4	F	F	${\sim}P \& {\sim}Q$

Each line on the truth-table can also be described by a kind of proposition called a **state-description**. A state-description in language $\mathcal{L}$ is a conjunction in which (1) each conjunct is either an atomic proposition of $\mathcal{L}$ or its negation; and (2) each atomic proposition of $\mathcal{L}$ appears exactly once. For example, P & Q and $\sim P$ & Q are each state-descriptions. A state-description succinctly describes the possible worlds associated with a line on the truth-table. For example, the possible worlds in region s_3 are just those in which P is false and Q is true; in other words, they are just those in which the state-description $\sim P$ & Q is true. Given any language, its state-descriptions will form a partition.[5]

Notice that the state-descriptions available for use are dependent on the language we are working with. If instead of language $\mathcal{L}$ we are working with a language $\mathcal{L}'$ containing three atomic propositions (P, Q, and R), we will have eight state-descriptions available instead of $\mathcal{L}$'s four. (You'll work with these eight state-descriptions in Exercise 2.1. For now we'll go back to working with language $\mathcal{L}$ and its paltry four.)

Every non-contradictory proposition in a language has an equivalent that is a disjunction of state-descriptions. We call this disjunction the proposition's **disjunctive normal form**. For example, the proposition $P \lor Q$ is true in regions s_1, s_2, and s_3. Thus

$$P \lor Q \dashv\vDash (P \,\&\, Q) \lor (P \,\&\, \sim Q) \lor (\sim P \,\&\, Q) \tag{2.1}$$

The proposition on the right-hand side is the disjunctive normal form equivalent of $P \lor Q$. To find the disjunctive normal form of a non-contradictory proposition, figure out which lines of the truth-table it's true on, then make a disjunction of the state-descriptions associated with each such line.[6]

2.1.3 Predicate logic

Sometimes we will want to work with languages that represent objects and properties. To do so, we first identify a **universe of discourse**, the total set of objects under discussion. Each object in the universe of discourse will be represented by a **constant**, which will usually be a lower-case letter ($a, b, c, \ldots$). Properties of those objects and relations among them will be represented by **predicates**, which will be capital letters.

Relations among propositions in such a language are exactly as described in the previous sections, except that we have two new kinds of propositions. First, our atomic propositions are now generated by applying a predicate to a constant, as in "Fa". Second, we can generate quantified sentences, as in "$(\forall x)(Fx \supset \sim Fx)$". Since we will rarely be using predicate logic, I won't work through the details here; a thorough treatment can be found in any introductory logic text.

I do want to emphasize, though, that as long as we restrict our attention to finite universes of discourse, all the logical relations we need can be handled by the propositional machinery discussed above. If, say, our only two constants are a and b and our only predicate is F, then the only atomic propositions in $\mathcal{L}$ will be Fa and Fb, for which we can build a standard truth-table:

Fa	Fb	state-description
T	T	Fa & Fb
T	F	Fa & $\sim Fb$
F	T	$\sim Fa$ & Fb
F	F	$\sim Fa$ & $\sim Fb$

For any proposition in this language containing a quantifier, we can find an equivalent composed entirely of atomic propositions and propositional connectives. To do this we need the notion of a **substitution instance**: a substitution instance of a quantified sentence is produced by removing the quantifier and replacing its variable throughout what remains with the same constant. (So, for example, $Fa \supset \sim Fa$ is a substitution instance of $(\forall x)(Fx \supset \sim Fx)$.) A universally quantified sentence is equivalent to a *conjunction* of all its substitution instances for constants in $\mathcal{L}$, while an existentially quantified sentence is equivalent to a *disjunction* of its substitution instances. For example, when our only two constants are a and b we have:

$$(\forall x)(Fx \supset \sim Fx) \; \dashv\vDash \; (Fa \supset \sim Fa) \; \& \; (Fb \supset \sim Fb) \tag{2.2}$$

$$(\exists x)Fx \; \dashv\vDash \; Fa \vee Fb \tag{2.3}$$

As long as we stick to finite universes of discourse, every proposition will have an equivalent that uses only propositional connectives. So even when we work in predicate logic, every non-contradictory proposition will have an equivalent in disjunctive normal form.

2.2 The probability axioms

A **distribution** over language $\mathcal{L}$ assigns a real number to each proposition in the language.[7] Bayesians represent an agent's degrees of belief as a distribution over a language; I will use "cr" to symbolize an agent's credence distribution.

For example, if an agent is 70% confident that it will rain tomorrow, we will write

$$cr(R) = 0.7 \qquad (2.4)$$

where R is the proposition that it will rain tomorrow. Another way to put this is that the agent's **unconditional credence** in rain tomorrow is 0.7. (*Unconditional* credences contrast with *conditional* credences, which we will discuss in Chapter 3.) The higher the numerical value of an agent's unconditional credence in a proposition, the more confident the agent is that that proposition is true.

Bayesians hold that a *rational* credence distribution satisfies certain rules. Among these are our first three core rules, **Kolmogorov's axioms**:

Non-Negativity: For any proposition P in $\mathcal{L}$, $cr(P) \geqslant 0$.

Normality: For any tautology T in $\mathcal{L}$, $cr(\mathsf{T}) = 1$.

Finite Additivity: For any mutually exclusive propositions P and Q in $\mathcal{L}$, $cr(P \vee Q) = cr(P) + cr(Q)$.

Kolmogorov's axioms are often referred to as "the probability axioms". Mathematicians call any distribution that satisfies these axioms a **probability distribution**. Kolmogorov (1933/1950) was the first to articulate these axioms as the foundation of mathematical probability theory.[8]

Warning

Kolmogorov's work inaugurated a mathematical field of probability theory distinct from the philosophical study of what probability is. While this was an important advance, it gave the word "probability" a special meaning in mathematical circles that can generate confusion elsewhere.

For a twenty-first-century mathematician, Kolmogorov's axioms *define* what it is for a distribution to be a "probability distribution". This is distinct from the way people use "probability" in everyday life. For one thing, the word "probability" in English may not mean the same thing in every use. And even if it does, it would be a substantive philosophical thesis that probabilities (in the everyday sense) can be represented by a numerical distribution satisfying Kolmogorov's axioms. Going in the other direction, there are numerical distributions satisfying the axioms that don't count as "probabilistic" in any ordinary sense. For example, we could invent a distribution "tv" that assigns 1 to every true proposition and 0 to every false proposition. To a mathematician, the fact that tv satisfies Kolmogorov's axioms makes it a probability distribution. But a proposition's tv-value might not match its probability in the everyday sense. Improbable propositions can turn out to be true (I just rolled snake-eyes!), and propositions with high probabilities can turn out to be false (the Titanic should've made it to port).

Probabilism is the philosophical view that rationality requires an agent's credences to form a probability distribution (that is, to satisfy Kolmogorov's axioms). Probabilism is attractive in part because it has intuitively appealing consequences. For example, from the probability axioms we can prove:

Negation: For any proposition P in $\mathcal{L}$, $cr(\sim P) = 1 - cr(P)$.

According to Negation, rationality requires an agent with $cr(R) = 0.7$ to have $cr(\sim R) = 0.3$. Among other things, Negation embodies the sensible thought that if you're highly confident that a proposition is true, you should be dubious that its negation is.

Usually I'll leave it as an exercise to prove that a particular consequence follows from the probability axioms, but here I will prove Negation as an example for the reader.

Negation Proof:

Claim	Justification
(1) P and $\sim P$ are mutually exclusive	logic
(2) $cr(P \vee \sim P) = cr(P) + cr(\sim P)$	(1), Finite Additivity
(3) $P \vee \sim P$ is a tautology	logic
(4) $cr(P \vee \sim P) = 1$	(3), Normality
(5) $1 = cr(P) + cr(\sim P)$	(2), (4)
(6) $cr(\sim P) = 1 - cr(P)$	(5), algebra

2.2.1 Consequences of the probability axioms

Below are a number of further consequences of the probability axioms. Again, these consequences are listed in part to illustrate intuitive things that follow from the axioms. But I'm also listing them because they'll be useful in future proofs.

Maximality: For any proposition P in $\mathcal{L}$, $\mathrm{cr}(P) \leqslant 1$.

Contradiction: For any contradiction F in $\mathcal{L}$, $\mathrm{cr}(\mathsf{F}) = 0$.

Entailment: For any propositions P and Q in $\mathcal{L}$, if $P \vDash Q$ then
$$\mathrm{cr}(P) \leqslant \mathrm{cr}(Q).$$

Equivalence: For any propositions P and Q in $\mathcal{L}$, if $P =\!\!\vDash Q$ then
$$\mathrm{cr}(P) = \mathrm{cr}(Q).$$

General Additivity: For any propositions P and Q in $\mathcal{L}$,
$$\mathrm{cr}(P \lor Q) = \mathrm{cr}(P) + \mathrm{cr}(Q) - \mathrm{cr}(P \,\&\, Q).$$

Finite Additivity (Extended):   For any finite set of mutually exclusive propositions $\{P_1, P_2, \ldots, P_n\}$,
$$\mathrm{cr}(P_1 \lor P_2 \lor \ldots \lor P_n) = \mathrm{cr}(P_1) + \mathrm{cr}(P_2) + \ldots + \mathrm{cr}(P_n).$$

Decomposition: For any propositions P and Q in $\mathcal{L}$,
$$\mathrm{cr}(P) = \mathrm{cr}(P \,\&\, Q) + \mathrm{cr}(P \,\&\, {\sim}Q).$$

Partition: For any finite partition of propositions in $\mathcal{L}$, the sum of their unconditional cr-values is 1.

Together, Non-Negativity and Maximality establish the bounds of our credence scale. Rational credences will always fall between 0 and 1 (inclusive). Given these bounds, Bayesians represent absolute certainty that a proposition is true as a credence of 1 and absolute certainty that a proposition is false as credence 0. The upper bound is arbitrary—we could have set it at whatever positive real number we wanted. But using 0 and 1 lines up nicely with everyday talk of being 0% confident or 100% confident in particular propositions, and also with various considerations of frequency and chance discussed later in this book.

Entailment is plausible for all the same reasons Comparative Entailment was plausible in Chapter 1; we've simply moved from an expression in terms of confidence comparisons to one using numerical credences. Understanding equivalence as mutual entailment, Entailment entails Equivalence. General Additivity is a generalization of Finite Additivity that allows us to calculate an agent's credence in any disjunction, whether the disjuncts are mutually exclusive or not. (When the disjuncts *are* mutually exclusive, their conjunction is a contradiction, the $\mathrm{cr}(P \,\&\, Q)$ term equals 0, and General Additivity takes us back to Finite Additivity.)

Finite Additivity (Extended) can be derived by repeatedly applying Finite Additivity. Begin with any finite set of mutually exclusive propositions $\{P_1, P_2, \ldots, P_n\}$. By Finite Additivity,

$$\mathrm{cr}(P_1 \vee P_2) = \mathrm{cr}(P_1) + \mathrm{cr}(P_2) \tag{2.5}$$

Logically, since P_1 and P_2 are each mutually exclusive with P_3, $P_1 \vee P_2$ is also mutually exclusive with P_3. So Finite Additivity yields

$$\mathrm{cr}([P_1 \vee P_2] \vee P_3) = \mathrm{cr}(P_1 \vee P_2) + \mathrm{cr}(P_3) \tag{2.6}$$

Combining Equations (2.5) and (2.6) then gives us

$$\mathrm{cr}(P_1 \vee P_2 \vee P_3) = \mathrm{cr}(P_1) + \mathrm{cr}(P_2) + \mathrm{cr}(P_3) \tag{2.7}$$

Next we would invoke the fact that $P_1 \vee P_2 \vee P_3$ is mutually exclusive with P_4 to derive

$$\mathrm{cr}(P_1 \vee P_2 \vee P_3 \vee P_4) = \mathrm{cr}(P_1) + \mathrm{cr}(P_2) + \mathrm{cr}(P_3) + \mathrm{cr}(P_4) \tag{2.8}$$

Clearly this process iterates as many times as we need to reach

$$\mathrm{cr}(P_1 \vee P_2 \vee \ldots \vee P_n) = \mathrm{cr}(P_1) + \mathrm{cr}(P_2) + \ldots + \mathrm{cr}(P_n) \tag{2.9}$$

The idea here is that once you have Finite Additivity for proposition sets of size two, you have it for proposition sets of any larger finite size as well. When the propositions in a finite set are mutually exclusive, the probability of their disjunction equals the sum of the probabilities of the disjuncts.

Combining Finite Additivity and Equivalence yields Decomposition. For any P and Q, P is equivalent to the disjunction of the mutually exclusive propositions $P \& Q$ and $P \& {\sim} Q$, so $\mathrm{cr}(P)$ must equal the sum of the cr-values of those two. Partition then takes a finite set of mutually exclusive propositions whose disjunction is a tautology. By Finite Additivity (Extended) the cr-values of the propositions in the partition must sum to the cr-value of the tautology, which by Normality must be 1.

2.2.2 A Bayesian approach to the Lottery scenario

In future sections I'll explain some alternative ways of thinking about the probability calculus. But first let's use probabilities to *do* something: a Bayesian analysis of the situation in the Lottery Paradox. Recall the scenario from Chapter 1: A fair lottery has one million tickets.[9] An agent is skeptical of each ticket that it will win, but takes it that some ticket will win. In Chapter 1 we saw that it's difficult to articulate plausible norms on binary belief that depict this agent as believing rationally. But once we move to degrees of belief, the analysis is straightforward.

We'll use a language in which the constants $a, b, c, \ldots$ stand for the various tickets in the lottery, and the predicate W says that a particular ticket wins. A reasonable credence distribution over the resulting language sets

$$cr(Wa) = cr(Wb) = cr(Wc) = \ldots = 1/1{,}000{,}000 \qquad (2.10)$$

Negation then gives us

$$cr(\sim Wa) = cr(\sim Wb) = cr(\sim Wc) = \ldots = 1 - 1/1{,}000{,}000 = 0.999999 \qquad (2.11)$$

reflecting the agent's high confidence for each ticket that that ticket won't win.

What about the disjunction saying that some ticket will win? Since the $Wa, Wb, Wc, \ldots$ propositions are mutually exclusive, Finite Additivity (Extended) yields

$$cr(Wa \lor Wb \lor Wc \lor Wd \lor \ldots) = \\ cr(Wa) + cr(Wb) + cr(Wc) + cr(Wd) + \ldots \qquad (2.12)$$

On the right-hand side of Equation (2.12) we have one million terms, each of which has a value of $1/1{,}000{,}000$. Thus the credence on the left-hand side equals 1.

The Lottery *Paradox* is a problem for particular norms on binary belief. We haven't done anything to resolve that paradox here. Instead, we've shown that the lottery situation giving rise to the paradox can be easily modeled by Bayesian means. We've build a model of the lottery situation in which the agent is both highly confident that some ticket will win and highly confident of each ticket that it will not. (Constructing a similar model for the Preface is left as an exercise for the reader.) There is no tension with the rules of rational confidence represented in Kolmogorov's axioms. The Bayesian model not only accommodates but *predicts* that if a rational agent has a small confidence in each of a set of mutually exclusive propositions, yet has a large enough number of those propositions available, that agent will be certain (or close to certain) that at least one of the propositions is true.

This analysis also reveals why it's difficult to simultaneously maintain both the Lockean thesis and the Belief Consistency norm from Chapter 1. The Lockean thesis implies that a rational agent believes a proposition just in case her credence in that proposition is above some numerical threshold. For any such threshold we pick (less than 1), it's possible to generate a lottery-type scenario in which the agent's credence that at least one ticket will win clears the threshold, but her credence for any given ticket that that ticket will lose also clears the threshold. Given the Lockean thesis, a rational agent will therefore believe that at least one ticket will win but also believe of each ticket that it will lose. This violates Belief Consistency, which says that every rational belief set is logically consistent.

2.2.3 Doxastic possibilities

In the previous section we considered propositions of the form Wx, each of which says of some particular ticket that it will win the lottery. To perform various calculations involving these W propositions, we assumed that they form a partition—that is, that they are mutually exclusive and jointly exhaustive. But you may worry that this isn't right: what about worlds in which ticket a and ticket b both win the lottery due to a clerical error, or worlds in which no ticket wins the lottery, or worlds in which the lottery never takes place, or worlds in which humans never evolve? These worlds are **logically possible**—the laws of logic alone don't rule them out. Yet the credence distribution we crafted for our agent assigns these worlds degree of belief 0. Could it ever be rational for an agent to assign a logical possibility *no* credence whatsoever?

We will refer to the possible worlds an agent entertains as her **doxastically possible worlds**.[10] Perhaps a fully rational agent never rules out any logically possible world; if so, then a rational agent's set of doxastic possibilities is always the full set of logical possibilities, and includes worlds like the one in which every ticket wins, the one in which no ticket wins, the one in which no humans exist, etc. We will discuss this position when we turn to the Regularity Principle in Chapters 4 and 5. For the time being I want to note that even if a rational agent should never entirely rule out a logically possible world, it might be convenient in particular contexts for her to *temporarily* ignore certain worlds as live possibilities. Pollsters calculating confidence intervals for their latest sampling data don't factor in the possibility that our sun will explode before the next presidential election.

How is the probability calculus affected when an agent restricts her doxastically possible worlds to a proper subset of the logically possible worlds? Section 2.1 defined various relations among propositions in terms of possible worlds. In that context, the appropriate set of possible worlds to consider was the full set of logically possible worlds. But we can reinterpret those definitions as quantified over an agent's doxastically possible worlds. In our analysis of the Lottery scenario above, we effectively ignored possible worlds in which no tickets win the lottery or in which more than one ticket wins. For our purposes it was simpler to suppose that the agent rules them out of consideration. So our Bayesian model treated each Wx proposition as mutually exclusive with all the others, allowing us to apply Finite Additivity to generate equations like (2.12). If we were working with the full space of logically possible worlds we would have worlds in which more than one Wx proposition was true, so those propositions wouldn't count as mutually exclusive. But relative to the set of possible worlds we've supposed the agent entertains, they are.

2.2.4 Probabilities are weird! The Conjunction Fallacy

As you work with credences it's important to remember that probabilistic relations can function very differently from the relations among categorical concepts that inform many of our intuitions. In the Lottery situation it's perfectly rational for an agent to be highly confident of a disjunction while having low confidence in each of its disjuncts. That may seem strange.

Tversky and Kahneman (1983) offer another probabilistic example that runs counter to most people's intuitions. In a famous study, they presented subjects with the following prompt:

> Linda is 31 years old, single, outspoken, and very bright. She majored in philosophy. As a student, she was deeply concerned with issues of discrimination and social justice, and also participated in anti-nuclear demonstrations.

The subjects were then asked to rank the probabilities of the following propositions (among others):

- Linda is active in the feminist movement.
- Linda is a bank teller.
- Linda is a bank teller and is active in the feminist movement.

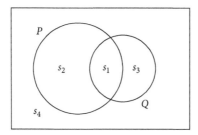

Figure 2.4 Areas equal to unconditional credences

The "great majority" of Tversky and Kahneman's subjects ranked the conjunction as more probable than the bank teller proposition. But this violates the probability axioms! A conjuction will always entail each of its conjuncts. By our Entailment rule—which follows from the probability axioms—the conjunct must be at least as probable as the conjunction. Being more confident in a conjunction than its conjunct is known as the **Conjunction Fallacy**.

2.3 Alternative representations of probability

2.3.1 Probabilities in Venn diagrams

Earlier we used Venn diagrams to visualize propositions and the relations among them. We can also use Venn diagrams to picture probability distributions. All we have to do is attach significance to something that was unimportant before: the *sizes* of regions in the diagram. We stipulate that the area of the entire rectangle is 1. The area of a region inside the rectangle equals the agent's unconditional credence in any proposition associated with that region. (Note that this visualization technique works only for credence distributions that satisfy the probability axioms.)[11]

For example, consider Figure 2.4. There we've depicted a probabilistic credence distribution in which the agent is more confident of proposition P than she is of proposition Q, as indicated by the P-circle's being larger than the Q-circle. What about $cr(Q \& P)$ versus $cr(Q \& {\sim}P)$? On the diagram the region labeled s_3 has slightly more area than the region labeled s_1, so the agent is slightly more confident of $Q \& {\sim}P$ than $Q \& P$. (When you construct your own Venn diagrams you need not include state-description labels like "s_3"; I've added them for reference.)

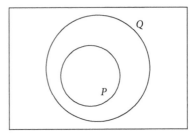

Figure 2.5 $P \vDash Q$

Warning

It's tempting to think that the size of a region in a Venn diagram represents the *number* of possible worlds in that region—the number of worlds that make the associated proposition true. But this would be a mistake. Just because an agent is more confident of one proposition than another does not necessarily mean she associates more possible worlds with the former than the latter. For example, if I tell you I have a weighted die that is more likely to come up six than any other number, your increased confidence in six does not necessarily mean that you think there are disproportionately many *worlds* in which the die lands six. The area of a region in a Venn diagram is a useful visual representation of an agent's confidence in its associated proposition. We should not read too much into it about the distribution of possible worlds.[12]

Venn diagrams make it easy to see why certain probabilistic relations hold. For example, take the General Additivity rule from Section 2.2.1. In Figure 2.4, the $P \vee Q$ region contains every point that is in the P-circle, in the Q-circle, or in both. We could calculate the area of that region by adding up the area of the P-circle and the area of the Q-circle, but in doing so we'd be counting the $P \& Q$ region (labeled s_1) twice. We adjust for this double-counting as follows:

$$\mathrm{cr}(P \vee Q) = \mathrm{cr}(P) + \mathrm{cr}(Q) - \mathrm{cr}(P \& Q) \tag{2.13}$$

That's General Additivity.

Figure 2.5 depicts a situation in which proposition P entails proposition Q. As discussed earlier, this requires the P-circle to be wholly contained within the

Q-circle. But since areas now represent unconditional credences, the diagram makes it obvious that the cr-value of proposition Q must be at least as great as the cr-value of proposition P. That's exactly what our Entailment rule requires. (It also shows why the Conjunction Fallacy is a mistake—imagine Q is the proposition that Linda is a bank teller and P is the proposition that Linda is a feminist bank teller.)

Venn diagrams can be a useful way of visualizing probability relationships. Bayesians often clarify a complex situation by sketching a quick Venn diagram of the agent's credence distribution. There are limits to this technique; when our languages grow beyond three or so atomic propositions it becomes difficult to get all the overlapping regions one needs and to make areas proportional to credences. But there are also cases in which it's much easier to understand why a particular theorem holds by looking at a diagram than by working with the axioms.

2.3.2 Probability tables

Besides being represented visually in a Venn diagram, a probability distribution can be represented precisely and efficiently in a **probability table**. To build a probability table, we begin with a set of propositions forming a partition of the agent's doxastic possibilities. For example, suppose an agent is going to roll a loaded six-sided die that comes up six on half of its rolls (with the remaining rolls distributed equally among the other numbers). A natural partition of the agent's doxastic space uses the propositions that the die comes up one, the die comes up two, the die comes up three, etc. The resulting probability table looks like this:

proposition	cr
Die comes up one.	1/10
Die comes up two.	1/10
Die comes up three.	1/10
Die comes up four.	1/10
Die comes up five.	1/10
Die comes up six.	1/2

The probability table first lists the propositions in the partition. Then for each proposition it lists the agent's unconditional credence in that proposition. If the agent's credences satisfy the probability axioms, the credence values in the table will satisfy two important constraints:

1. Each value is non-negative.
2. The values in the column sum to 1.

The first rule follows from Non-Negativity, while the second follows from our Partition theorem.

Once we know the credences of partition members, we can calculate the agent's unconditional credence in any other proposition expressible in terms of that partition. First, any contradiction receives credence 0. Then for any other proposition, we figure out which rows of the table it's true on, and calculate its credence by summing the values on those rows. For example, we might be interested in the agent's credence that the die roll comes up even. The proposition that the roll comes up even is true on the second, fourth, and sixth rows of the table. So the agent's credence in that proposition is $1/10 + 1/10 + 1/2 = 7/10$.

We can calculate the agent's credence in this way because

$$E \dashv\vdash 2 \vee 4 \vee 6 \tag{2.14}$$

where E is the proposition that the die came up even, "2" represents its coming up two, etc. By Equivalence,

$$cr(E) = cr(2 \vee 4 \vee 6) \tag{2.15}$$

Since the propositions on the right are members of a partition, they are mutually exclusive, so Finite Additivity (Extended) yields

$$cr(E) = cr(2) + cr(4) + cr(6) \tag{2.16}$$

The agent's unconditional credence in E can be found by summing the values on the second, fourth, and sixth rows of the table.

Given a propositional language $\mathcal{L}$, it's often useful to build a probability table using the partition containing $\mathcal{L}$'s state-descriptions. For example, for a language with two atomic propositions P and Q, I might give you the following probability table:

	P	Q	cr
s_1	T	T	0.1
s_2	T	F	0.3
s_3	F	T	0.2
s_4	F	F	0.4

The state-descriptions in this table are fully specified by the Ts and Fs appearing under P and Q in each row, but I've also provided labels $(s_1, s_2, \ldots)$ for each state-description to show how they correspond to regions in Figure 2.4.

Suppose a probabilistic agent has the unconditional credences specified in this table. What credence does she assign to $P \vee Q$? From the Venn diagram we can see that $P \vee Q$ is true on state-descriptions s_1, s_2, and s_3. So we find $cr(P \vee Q)$ by adding up the cr-values on the first three rows of our table. In this case $cr(P \vee Q) = 0.6$.

A probability table over state-descriptions is a particularly efficient way of specifying an agent's unconditional credence distribution over an entire propositional language.[13] A language $\mathcal{L}$ closed under the standard connectives contains infinitely many propositions, so a distribution over that language contains infinitely many values. If the agent's credences satisfy the probability axioms, the Equivalence rule tells us that equivalent propositions must all receive the same credence. So we can specify the entire distribution just by specifying its values over a maximal set of non-equivalent propositions in the language.

But that can still be a lot of propositions! If $\mathcal{L}$ has n atomic propositions, it will contain 2^{2^n} non-equivalent propositions (see Exercise 2.3). For 2 atomics that's only 16 credence values to specify, but by the time we reach 4 atomics it's up to 65,536 distinct values.

On the other hand, a language with n atomics will contain only 2^n state-descriptions. And once we provide unconditional credences for these propositions in our probability table, all the remaining values in the distribution follow. Every contradictory proposition receives credence 0, while each non-contradictory proposition is equivalent to a disjunction of state-descriptions (its disjunctive normal form). By Finite Additivity (Extended), the credence in a disjunction of state-descriptions is just the sum of the credences assigned to those state-descriptions. So the probability table contains all the information we need to specify the full distribution.[14]

2.3.3 Using probability tables

Probability tables describe an entire credence distribution in an efficient manner; instead of specifying a credence value for each non-equivalent proposition in the language, we need only specify values for its state-descriptions. Credences in state-descriptions can then be used to calculate credences in other propositions.

But probability tables can also be used to prove theorems and solve problems. To do so, we replace the numerical credence values in the table with variables:

	P	Q	cr
s_1	T	T	a
s_2	T	F	b
s_3	F	T	c
s_4	F	F	d

This probability table for an $\mathcal{L}$ with two atomic propositions makes no assumptions about the agent's specific credence values. It is therefore fully general, and can be used to prove general theorems about probability distributions. For example, on this table

$$\text{cr}(P) = a + b \tag{2.17}$$

But a is just $\text{cr}(P \,\&\, Q)$, and b is $\text{cr}(P \,\&\, \sim Q)$. This gives us a very quick proof of the Decomposition rule from Section 2.2.1. It's often much easier to prove a general probability result using a probability table built on state-descriptions than it is to prove the same result from Kolmogorov's axioms.

As for problem-solving, suppose I tell you that my credence distribution satisfies the probability axioms and also has the following features: I am certain of $P \vee Q$, and I am equally confident in Q and $\sim Q$. I then ask you to tell me my credence in $P \supset Q$.

You might be able to solve this problem by drawing a careful Venn diagram—perhaps you can even solve it in your head! If not, the probability table provides a purely algebraic solution method. We start by expressing the constraints on my distribution as equations using the variables from the table. From our second constraint on probability tables we have:

$$a + b + c + d = 1 \tag{2.18}$$

(Sometimes it also helps to invoke the first constraint, writing inequalities specifying that $a, b, c,$ and d are each greater than or equal to 0. In this particular problem those inequalities aren't needed.) Next we represent the fact that I am equally confident in Q and $\sim Q$:

$$\text{cr}(Q) = \text{cr}(\sim Q) \tag{2.19}$$
$$a + c = b + d \tag{2.20}$$

Finally, we represent the fact that I am certain of $P \lor Q$. The only line of the table on which $P \lor Q$ is false is line s_4; if I'm certain of $P \lor Q$, I must assign this state-description a credence of 0. So

$$d = 0 \tag{2.21}$$

Now what value are we looking for? I've asked you for my credence in $P \supset Q$; that proposition is true on lines s_1, s_3, and s_4; so you need to find $a + c + d$. Applying a bit of algebra to Equations (2.18), (2.20), and (2.21), you should be able to determine that $a + c + d = 1/2$.

2.3.4 Odds

Agents sometimes report their levels of confidence using odds rather than probabilities. If an agent's unconditional credence in P is $\mathrm{cr}(P)$, her **odds for** P are $\mathrm{cr}(P) : \mathrm{cr}(\sim P)$, and her **odds against** P are $\mathrm{cr}(\sim P) : \mathrm{cr}(P)$.

For example, there are thirty-seven pockets on a European roulette wheel. (American wheels have more.) Eighteen of those pockets are black. Suppose an agent's credences obey the probability axioms, and she assigns equal credence to the roulette ball's landing in any of the thirty-seven pockets. Then her credence that the ball will land in a black pocket is 18/37, and her credence that it won't is 19/37. Her odds for black are therefore

$$18/37 : 19/37, \text{ or } 18 : 19 \tag{2.22}$$

(Since the agent assigns equal credence to each of the pockets, these odds are easily found by comparing the number of pockets that make the proposition true to the number of pockets that make it false.) Yet in gambling contexts we usually report odds *against* a proposition. So in a casino someone might say that the odds against the ball's landing in the single green pocket are "36 to 1". The odds against an event are tightly connected to the stakes at which it would be fair to gamble on that event, which we will discuss in Chapter 7.

Warning

Instead of using a colon or the word "to", people sometimes quote odds as fractions. So someone might say that the odds for the roulette ball's landing in a black pocket are "18/19".[15] It's important not to mistake this fraction for a probability value. If your odds for black are 18 : 19, you take the ball's landing on black to a bit less likely to happen than not. But if your unconditional credence in black were 18/19, you would always bet on black!

It can be useful to think in terms of odds not only for calculating betting stakes but also because odds highlight differences that may be obscured by probability values. Suppose you hold a single ticket in a lottery that you take to be fair. Initially you think that the lottery contains only two tickets, of which yours is one. But then someone tells you there are 100 tickets in the lottery. This is a significant blow to your chances, witnessed by the fact that your assessment of the odds against winning has gone from 1 : 1 to 99 : 1. The significance of this change can also be seen in your unconditional credence that you will lose, which has jumped from 50% to 99%.

But now it turns out that your informant was misled, and there are actually 10,000 tickets in the lottery! This is another significant blow to your chances, intuitively at least as bad as the first jump in size. And indeed, your odds against winning go from 99 : 1 to 9, 999 : 1. Yet your credence that you'll lose moves only from 99% to 99.99%. Probabilities work on an additive scale; from that perspective a move from 0.5 to 0.99 looks important while a move from 0.99 to 0.9999 looks like a rounding error. But odds use ratios, which highlight multiplicative effects more obviously.

2.4 What the probability calculus adds

In Chapter 1 we moved from thinking of agents' doxastic attitudes in terms of binary (categorical) beliefs and confidence comparisons to working with numerical degrees of belief. At a first pass, this is a purely *descriptive* maneuver, yielding descriptions of an agent's attitudes at a higher fineness of grain. As we saw in Chapter 1, this added level of descriptive detail confers both advantages and disadvantages. On the one hand, credences allow us to say *how much more* confident an agent is of one proposition than another. On the other hand, assigning numerical credences over a set of propositions introduces a complete ranking, making all the propositions commensurable with respect to the agent's confidences. This may be an unrealistic result.

Chapter 1 also offered a *norm* for comparative confidence rankings:

Comparative Entailment: For any pair of propositions such that the first entails the second, rationality requires an agent to be at least as confident of the second as the first.

We have now introduced Kolmogorov's probability axioms as a set of norms on credences. Besides the descriptive changes that happen when we move from comparative confidences to numerical credences, how do the probability axioms go beyond Comparative Entailment? What *more* do we demand of an agent when we require that her credences be probabilistic?

Comparative Entailment can be derived from the probability axioms—we've already seen that by the Entailment rule, if $P \vDash Q$ then rationality requires $cr(P) \leq cr(Q)$. But how much of the probability calculus can be recreated simply by assuming that Comparative Entailment holds? We saw in Chapter 1 that if Comparative Entailment holds, a rational agent will assign equal, maximal confidence to all tautologies and equal, minimal confidence to all contradictions. This doesn't assign specific *numerical confidence values* to contradictions and tautologies, because Comparative Entailment doesn't work with numbers. But the probability axioms' 0-to-1 scale for credence values is fairly stipulative and arbitrary anyway. The real essence of Normality, Contradiction, Non-Negativity, and Maximality can be obtained from Comparative Entailment.

That leaves one axiom unaccounted for. To me the key insight of probabilism—and the element most responsible for Bayesianism's distinctive contributions to epistemology—is Finite Additivity. Finite Additivity places demands on rational credence that don't follow from any of the comparative norms we've seen. To see how, consider the following two credence distributions over a language with one atomic proposition:

Mr. Prob:	$cr(F) = 0$	$cr(P) = 1/6$	$cr(\sim P) = 5/6$	$cr(T) = 1$
Mr. Weak:	$cr(F) = 0$	$cr(P) = 1/36$	$cr(\sim P) = 25/36$	$cr(T) = 1$

With respect to their confidence comparisons, Mr. Prob and Mr. Weak are identical; they each rank $\sim P$ above P and both those propositions between a tautology and a contradiction. Both agents satisfy Comparative Entailment. Both agents also satisfy the Non-Negativity and Normality probability axioms. But only Mr. Prob satisfies Finite Additivity. His credence in the tautologous disjunction $P \lor \sim P$ is the sum of his credences in its mutually exclusive disjuncts. Mr. Weak's credences, on the other hand, are **superadditive**: he assigns *more* credence to the disjunction than the sum of his credences in its mutually exclusive disjuncts ($1 > 1/36 + 25/36$).

Probabilism goes beyond Comparative Entailment by exalting Mr. Prob over Mr. Weak. In endorsing Finite Additivity, the probabilist holds that Mr. Weak's credences have an *irrational* feature not present in Mr. Prob's. When we apply Bayesianism in later chapters, we'll see that Finite Additivity gives rise to some of the theory's most interesting and useful results. It does so by demanding that rational credences be *linear*, in the sense that a disjunction's credence is a linear combination[16] of the credences in its mutually exclusive disjuncts.

Of course, the fan of confidence comparisons need not restrict herself to the Comparative Entailment norm. Chapter 14 will explore further comparative constraints that have been proposed, some of which are capable of discriminating between Mr. Prob and Mr. Weak. We will ask whether those non-numerical norms can replicate all the desirable results secured by Finite Additivity for the Bayesian credence regime. This will be an especially pressing question because the impressive Bayesian numerical results come with a price. When we examine explicit philosophical arguments for the probability axioms in Part IV of this book, we'll find that while Normality and Non-Negativity can be straightforwardly argued for, Finite Additivity is the most difficult part of Bayesian epistemology to defend.

2.5 Exercises

Problem 2.1. 🎵
(a) List all eight state-descriptions available in a language with the three atomic sentences P, Q, and R.
(b) Give the disjunctive normal form of $(P \vee Q) \supset R$.

Problem 2.2. Here's a fact: For any non-contradictory propositions X and Y, $X \vDash Y$ if and only if every disjunct in the disjunctive normal form equivalent of X is also a disjunct of the disjunctive normal form equivalent of Y.
(a) 🎵 Use this fact to show that $(P \vee Q) \& R \vDash (P \vee Q) \supset R$.
(b) 🎵🎵🎵 Explain why the fact is true. (Be sure to explain both the "if" direction and the "only if" direction!)

Problem 2.3. 🎵🎵 Explain why a language $\mathcal{L}$ with n atomic propositions can express exactly 2^{2^n} non-equivalent propositions. (Hint: Think about the number of state-descriptions available, and the number of distinct disjunctive normal forms.)

Problem 2.4. 𝄐 Suppose your universe of discourse contains only two objects, named by the constants "a" and "b".
  (a) Find a quantifier-free equivalent of the proposition $(\forall x)[Fx \supset (\exists y)Gy]$.
  (b) Find the disjunctive normal form of your quantifier-free proposition from part (a).

Problem 2.5. 𝄐 Can a probabilistic credence distribution assign $cr(P) = 0.5$, $cr(Q) = 0.5$, and $cr(\sim P \& \sim Q) = 0.8$? Explain why or why not.[17]

Problem 2.6. 𝄐 Starting with only the probability axioms and Negation, write out proofs for all of the probability rules listed in Section 2.2.1. Your proofs must be straight from the axioms—no using Venn diagrams or probability tables! Once you prove a rule you may use it in further proofs. (Hint: You may want to prove them in an order different from the one in which they're listed. And I did Finite Additivity (Extended) for you.)

Problem 2.7. 𝄐 Prove that for any propositions P and Q, if $cr(P \equiv Q) = 1$ then $cr(P) = cr(Q)$.

Problem 2.8. 𝄐 In *The Empire Strikes Back*, C-3PO tells Han Solo that the odds against successfully navigating an asteroid field are 3,720 to 1.
  (a) What is C-3PO's unconditional credence that they will successfully navigate the asteroid field? (Express your answer as a fraction.)
  (b) Suppose C-3PO is certain that they will survive if they either successfully navigate the asteroid field, or fail to successfully navigate it but hide in a cave. He is also certain that those are the only two ways they can survive, and his odds against the conjunction of failing to successfully navigate and hiding in a cave are 59 to 2. Assuming C-3PO's credences obey the probability axioms, what are his odds against their surviving?
  (c) In the movie, how does Han respond to 3PO's odds declaration? (Apparently Han prefers his probabilities quoted as percentages.)

Problem 2.9. 𝄐 Consider the probabilistic credence distribution specified by this probability table:

P	Q	R	cr
T	T	T	0.1
T	T	F	0.2
T	F	T	0
T	F	F	0.3
F	T	T	0.1
F	T	F	0.2
F	F	T	0
F	F	F	0.1

Calculate each of the following values on this distribution:

(a) $cr(P \equiv Q)$
(b) $cr(R \supset Q)$
(c) $cr(P \& R) - cr(\sim P \& R)$
(d) $cr(P \& Q \& R)/cr(R)$

Problem 2.10. 🎶 Can an agent have a probabilistic cr-distribution meeting all of the following constraints?

1. The agent is certain of $A \supset (B \equiv C)$.
2. The agent is equally confident of B and $\sim B$.
3. The agent is twice as confident of C as $C \& A$.
4. $cr(B \& C \& \sim A) = 1/5$.

If not, prove that it's impossible. If so, provide a probability table and demonstrate that the resulting distribution satisfies each of the four constraints. (Hint: Start by building a probability table; then figure out what each of the constraints says about the credence values in the table; then figure out if it's possible to meet all of the constraints at once.)

Problem 2.11. 🖊 Tversky and Kahneman's finding that ordinary subjects commit the Conjunction Fallacy has held up to a great deal of experimental replication. Kolmogorov's axioms make it clear that the propositions involved cannot range from most probable to least probable in the way subjects consistently rank them. Do you have any suggestions for *why* subjects might consistently make this mistake? Is there any way to read what the subjects are doing as rationally acceptable?

Problem 2.12. 🖋 Recall Mr. Prob and Mr. Weak from Section 2.4. Mr. Weak assigns lower credences to each contingent proposition than does Mr. Prob.

While Mr. Weak's distribution satisfies Non-Negativity and Normality, it violates Finite Additivity by being superadditive: it contains a disjunction whose credence is *greater* than the sum of the credences of its mutually exclusive disjuncts.

Construct a credence distribution for "Mr. Bold" over language $\mathcal{L}$ with single atomic proposition P. Mr. Bold should rank every proposition in the same order as Mr. Prob and Mr. Weak. Mr. Bold should also satisfy Non-Negativity and Normality. But Mr. Bold's distribution should be **subadditive**: it should contain a disjunction whose credence is *less* than the sum of the credences of its mutually exclusive disjuncts.

2.6 Further reading

INTRODUCTIONS AND OVERVIEWS

Merrie Bergmann, James Moor, and Jack Nelson (2013). *The Logic Book*. 6th edition. New York: McGraw Hill

One of many available texts that thoroughly covers the logical material assumed in this book.

Ian Hacking (2001). *An Introduction to Probability and Inductive Logic*. Cambridge: Cambridge University Press
Brian Skyrms (2000). *Choice and Chance: An Introduction to Inductive Logic*. 4th edition. Stamford, CT: Wadsworth

Each of these books contains a Chapter 6 offering an entry-level, intuitive discussion of the probability rules—though neither explicitly uses Kolmogorov's axioms. Hacking has especially nice applications of probabilistic reasoning, along with many counterintuitive examples like the Conjunction Fallacy from our Section 2.2.4.

CLASSIC TEXTS

A. N. Kolmogorov (1933/1950). *Foundations of the Theory of Probability*. Translation edited by Nathan Morrison. New York: Chelsea Publishing Company

Text in which Kolmogorov laid out his famous axiomatization of probability theory.

Extended Discussion

> J. Robert G. Williams (2016). Probability and Non-Classical Logic. In: *Oxford Handbook of Probability and Philosophy*. Ed. by Alan Hájek and Christopher R. Hitchcock. Oxford: Oxford University Press

Covers probability distributions in non-classical logics, such as logics with non-classical entailment rules and logics with more than one truth-value. Also briefly discusses probability distributions in logics with extra connectives and operators, such as modal logics.

> Branden Fitelson (2008). A Decision Procedure for Probability Calculus with Applications. *The Review of Symbolic Logic* 1, pp. 111–125

Fills in the technical details of solving probability problems algebraically using probability tables (which Fitelson calls "stochastic truth-tables"), including the relevant meta-theory. Also describes a Mathematica package that will solve probability problems and evaluate probabilistic conjectures for you, downloadable for free at http://fitelson.org/PrSAT/.

Notes

1. Other authors describe degrees of belief as assigned to sentences, statements, or sets of events. Also, propositions are sometimes taken to be identical to one of these alternatives. As mentioned in Chapter 1, I will not assume much about what propositions are, except that: they are capable of having truth-values (that is, capable of being true or false); they are expressible by declarative sentences; and they have enough internal structure to contain logical operators. This last assumption could be lifted with a bit of work.
2. Bayesians sometimes define degrees of belief over a **sigma algebra**. A sigma algebra is a set of sets that is closed under (countable) union, (countable) intersection, and complementation. Given a language $\mathcal{L}$, the sets of possible worlds associated with the propositions in that language form a sigma algebra. The algebra is closed under union, intersection, and complementation because the propositions in $\mathcal{L}$ are closed under disjunction, conjunction, and negation (respectively).
3. I'm also going to be fairly cavalier about corner-quotes, the use-mention distinction, etc.

4. Throughout this book we will be assuming a classical logic, in which each proposition has exactly one of two available truth-values (true/false) and entailment obeys the inference rules taught in standard introductory logic classes. For information about probability in non-classical logics, see the Further Reading at the end of this chapter.

5. The cognoscenti will note that in order for the state-descriptions of $\mathcal{L}$ to form a partition, the atomic propositions of $\mathcal{L}$ must be (logically) independent. We will assume throughout this book that every propositional language employed contains logically independent atomic propositions, unless explicitly noted otherwise.

6. Strictly, in order to get the result that the state-descriptions in a language form a partition and the result that each non-contradictory proposition has a *unique* disjunctive normal form, we need to further regiment our definitions. To our definition of a state-description we add that the atomic propositions must appear in alphabetical order. We then introduce a canonical ordering of the state-descriptions in a language (say, the order in which they appear in a standardly ordered truth-table) and require disjunctive normal form propositions to contain their disjuncts in canonical order with no repetition.

7. In the statistics community, probability distributions are often assigned over the possible values of sets of random variables. Propositions are then thought of as dichotomous random variables capable of taking only the values 1 and 0 (for "true" and "false", respectively). Only rarely in this book will we look past distributions over propositions to distributions over more general random variables.

8. The axioms I've presented are not precisely identical to Kolmogorov's, but the differences are insignificant for our purposes. Some authors also include Countable Additivity—which we'll discuss in Chapter 5—among "Kolmogorov's axioms", but I'll use the phrase to pick out only Non-Negativity, Normality, and Finite Additivity.

   Galavotti (2005, pp. 54–5) notes that authors such as Mazurkiewicz (1932) and Popper (1938) also provided axioms for probability around the time Kolmogorov was working. She recommends Roeper and Leblanc (1999) for an extensive survey of the axiomatizations available.

9. This analysis could easily be generalized to any large, finite number of tickets.

10. Philosophers sometimes describe the worlds an agent entertains as her "epistemically possible worlds". Yet that term also carries a connotation of being determined by what the agent *knows*. So I'll discuss doxastically possible worlds, which are determined by what an agent *takes* to be possible rather than what she *knows*.

11. A probability distribution over sets of possible worlds is an example of what mathematicians call a "measure". The function that takes any region of a Euclidean two-dimensional space and outputs its area is also a measure. That makes probabilities representable by areas in a rectangle.

12. To avoid the confusion discussed here, some authors use "muddy" Venn diagrams in which all atomic propositions are associated with regions of the same size, and probability weights are indicated by piling up more or less "mud" on top of particular regions. Muddy Venn diagrams are difficult to depict on two-dimensional paper, so I've stuck with representing higher confidence as greater area.

13. Truth-tables famously come to us from Wittgenstein's *Tractatus Logico-Philosophicus* (Wittgenstein 1921/1961), in which Wittgenstein also proposed a theory of probability

assigning equal value to each state-description. But to my knowledge the first person to characterize probability distributions in *general* by the values they assign to state-descriptions was Carnap, as in his (1945, Sect. 3).

14. We have argued *from* the assumption that an agent's credences satisfy the probability axioms *to* the conclusion that her unconditional credence in any non-contradictory proposition is the sum of her credences in the disjuncts of its disjunctive normal form. One can also argue in the other direction. Suppose I stipulate an agent's credence distribution over language $\mathcal{L}$ as follows: (1) I stipulate unconditional credences for $\mathcal{L}$'s state-descriptions that are non-negative and sum to 1; (2) I stipulate that for every other non-contradictory proposition in $\mathcal{L}$, the agent's credence in that proposition is the sum of her credences in the disjuncts of that proposition's disjunctive normal form; and (3) I stipulate that the agent's credence in each contradiction is 0. We can prove that any credence distribution stipulated in this fashion will satisfy Kolmogorov's three probability axioms. I'll leave the (somewhat challenging) proof as an exercise for the reader.

15. Odds against a proposition, quoted with a slash like a fraction, are known as "British odds", for their popularity among British and Irish bookies.

16. Given two variables x and y and two constants a and b, we call $z = ax + by$ a **linear combination** of x and y. Finite Additivity makes $cr(X \vee Y)$ a linear combination of $cr(X)$ and $cr(Y)$, with the constants a and b each set to 1.

17. I owe this problem to Julia Staffel.

3

Conditional Credences

Chapter 2's discussion was confined to unconditional credence, an agent's outright degree of confidence that a particular proposition is true. This chapter takes up conditional credence, an agent's credence that one proposition is true on the supposition that another one is.

The main focus of this chapter is our fourth core normative Bayesian rule: the Ratio Formula. This rational constraint on conditional credences has a number of important consequences, including Bayes's Theorem (which gives Bayesianism its name).

Conditional credences are also central to the way Bayesians understand evidential relevance. I will define relevance as positive correlation, then explain how this notion has been used to investigate causal relations through the concept of screening off.

Having achieved a deeper understanding of the mathematics of conditional credences, I return at the end of the chapter to what exactly a conditional credence is. In particular, I discuss an argument by David Lewis that a conditional credence can't be understood as an unconditional credence in a conditional.

3.1 Conditional credences and the Ratio Formula

Arturo and Baxter know that two events will occur simultaneously in separate rooms: a fair coin will be flipped, and a clairvoyant will predict how it will land. Let H represent the proposition that the coin comes up heads, and C represent the proposition that the clairvoyant predicts heads. Suppose Arturo and Baxter each assign an unconditional credence of $1/2$ to H and an unconditional credence of $1/2$ to C.

Although Arturo and Baxter assign the same unconditional credences as each other to H and C, they still might take these propositions to be *related* in different ways. We could tease out those differences by saying to each agent, "I have no idea how the coin is going to come up or what the clairvoyant is going to say. But suppose for a moment the clairvoyant predicts heads. On this supposition, how confident are you that the coin will come up heads?" If

Fundamentals of Bayesian Epistemology 1: Introducing Credences. Michael G. Titelbaum, Oxford University Press.
© Michael G. Titelbaum 2022. DOI: 10.1093/oso/9780198707608.003.0003

Arturo says 1/2 and Baxter says 99/100, that's a good indication that Baxter has more faith in the mystical than Arturo.

The quoted question in the previous paragraph elicits Arturo and Baxter's *conditional* credences, as opposed to the *unconditional* credences discussed in Chapter 2. An unconditional credence is a degree of belief assigned to a single proposition, indicating how confident the agent is that that proposition is true. A **conditional credence** is a degree of belief assigned to an ordered pair of propositions, indicating how confident the agent is that the first proposition is true on the supposition that the second is. We symbolize conditional credences as follows:

$$cr(H \mid C) = 1/2 \tag{3.1}$$

This equation says that a particular agent (in this case, Arturo) has a 1/2 credence that the coin comes up heads conditional on the supposition that the clairvoyant predicts heads. The vertical bar indicates a conditional credence; to the right of the bar is the proposition supposed; to the left of the bar is the proposition evaluated in light of that supposition. The proposition to the right of the bar is sometimes called the **condition**; I am not aware of any generally accepted name for the proposition on the left.

To be clear: A real agent never assigns any credences *ex nihilo*, without assuming at least some background information. An agent's unconditional credences in various propositions (such as *H*) are informed by her background information at that time. To assign a conditional credence, the agent combines her stock of background information with a *further* supposition that the condition is true. She then evaluates the other proposition in light of this combination.

A conditional credence is assigned to an *ordered* pair of propositions. It makes a difference which proposition is supposed and which is evaluated. Consider a case in which I'm going to roll a fair die and you have various credences involving the proposition *E* that it comes up even and the proposition 6 that it comes up six. Compare:

$$cr(6 \mid E) = 1/3 \tag{3.2}$$
$$cr(E \mid 6) = 1 \tag{3.3}$$

3.1.1 The Ratio Formula

Section 2.2 described Kolmogorov's probability axioms, which Bayesians take to represent rational constraints on an agent's unconditional credences.

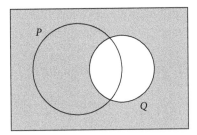

Figure 3.1 The region that dictates cr($P \mid Q$)

Bayesians then add a constraint relating conditional to unconditional credences:

Ratio Formula: For any P and Q in $\mathcal{L}$, if cr(Q) > 0 then

$$\text{cr}(P \mid Q) = \frac{\text{cr}(P \,\&\, Q)}{\text{cr}(Q)}$$

Stated this way, the Ratio Formula remains silent on the value of cr($P \mid Q$) when cr(Q) = 0. There are various positions on how one should assign conditional credences when the condition has credence 0; we'll address some of them in our discussion of the infinite in Section 5.4.

Why should an agent's conditional credence equal the ratio of those unconditionals? Consider Figure 3.1. The rectangle represents all the possible worlds the agent entertains. The agent's unconditional credence in P is the fraction of that rectangle taken up by the P-circle. (The area of the rectangle is stipulated to be 1, so that fraction is the area of the P-circle divided by 1, which is just the area of the P-circle.) When we ask the agent to assign a credence conditional on the supposition that Q, she temporarily narrows her focus to just those possibilities that make Q true. In other words, she excludes from her attention the worlds I've shaded in the diagram, and considers only what's in the Q-circle. The agent's credence in P conditional on Q is the fraction of the Q-circle occupied by P-worlds. So it's the area of the PQ overlap divided by the area of the entire Q-circle, which is cr($P \,\&\, Q$)/cr(Q).

In the scenario in which I roll a fair die, your initial doxastic possibilities include all six outcomes of the die roll. I then ask for your credence that the die comes up six conditional on its coming up even—that is, cr(6 $\mid E$). To assign this value, you exclude from consideration all the odd outcomes. You haven't actually *learned* that the die outcome is even; I've simply asked you to *suppose* that it comes up even and assign a confidence to other propositions in light of that supposition. You distribute your credence equally over the outcomes that remain under consideration (2, 4, and 6), so your credence in six conditional on even is 1/3.

We get the same result from the Ratio Formula:

$$cr(6 \mid E) = \frac{cr(6 \ \& \ E)}{cr(E)} = \frac{1/6}{1/2} = \frac{1}{3} \tag{3.4}$$

The Ratio Formula allows us to calculate your conditional credences (confidences under a supposition) from your unconditional credences (confidences relative to no suppositions beyond your background information). Hopefully it's obvious why E gets an unconditional credence of 1/2 in this case; as for 6 & E, that's equivalent to just 6, so it gets an unconditional credence of 1/6.[1]

Warning

Mathematicians often treat the Ratio Formula as a *definition* of conditional probability. From their point of view, a conditional probability has the value it does *in virtue of* two unconditional probabilities' standing in a certain ratio. But I do not want to reduce the possession of a conditional credence to the possession of two unconditional credences standing in a particular relation. I take a conditional credence to be a genuine mental state (an attitude toward an ordered pair of propositions) capable of being elicited in various ways, such as by asking an agent her confidence in a proposition given a supposition. So I will interpret the Ratio Formula as a rational constraint on how an agent's conditional credences should relate to her unconditional credences. As a normative *constraint* (rather than a *definition*), it can be violated—by assigning a conditional credence that doesn't equal the specified ratio.

The point of the previous warning is that the Ratio Formula is a rational constraint, and agents don't always meet all the rational constraints on their credences. Yet for agents who do satisfy the Ratio Formula, there can be no difference between their conditional credences without some difference in their unconditional credences as well. If we're both rational and I assign a different $cr(P \mid Q)$ value than you, we cannot assign the same values to both $cr(P \ \& \ Q)$ and $cr(Q)$. (A rational agent's conditional credences **supervene** on her unconditional credences.) Fully specifying a rational agent's unconditional credence distribution suffices to specify her conditional credences as well.[2] For instance, we might specify Arturo's and Baxter's credence distributions using the following probability table:

C	H	cr_A	cr_B
T	T	1/4	99/200
T	F	1/4	1/200
F	T	1/4	1/200
F	F	1/4	99/200

Here cr_A represents Arturo's credences and cr_B represents Baxter's. Arturo's unconditional credence in C is identical to Baxter's—the values on the first two rows sum to 1/2 for each of them. Similarly, Arturo and Baxter have the same unconditional credence in H (the sum of the first and third rows). Yet Arturo and Baxter disagree in their confidence that the coin will come up heads (H) given that the clairvoyant predicts heads (C). Using the Ratio Formula, we calculate this conditional credence by dividing the value on the first row of the table by the sum of the values on the first two rows. This yields:

$$cr_A(H \mid C) = \frac{1/4}{1/2} = \frac{1}{2} \neq \frac{99}{100} = \frac{99/200}{100/200} = cr_B(H \mid C) \qquad (3.5)$$

Baxter has high confidence in the clairvoyant's abilities. So on the supposition that the clairvoyant predicts heads, Baxter is almost certain that the flip comes up heads. Arturo, on the other hand, is skeptical, so supposing that the clairvoyant predicts heads leaves his opinions about the flip outcome unchanged.

3.1.2 Consequences of the Ratio Formula

Combining the Ratio Formula with the probability axioms yields further useful probability rules. First we have the

Law of Total Probability: For any proposition P and finite partition $\{Q_1, Q_2, \ldots, Q_n\}$ in $\mathcal{L}$,

$$cr(P) = cr(P \mid Q_1) \cdot cr(Q_1) + cr(P \mid Q_2) \cdot cr(Q_2) +$$
$$\ldots + cr(P \mid Q_n) \cdot cr(Q_n)$$

Suppose you're trying to predict whether I will bike to work tomorrow, but you're unsure if the weather will rain, hail, or be clear. The Law of Total Probability allows you to systematically work through the possibilities in that partition. You multiply your confidence that it will rain by your confidence that I'll bike should it rain. Then you multiply your confidence that it'll hail by your confidence in my biking given hail. Finally you multiply your unconditional credence that it'll be clear by your conditional credence that I'll bike given that it's clear. Adding these three products together yields your unconditional credence that I'll bike. (In the formula, the proposition that I'll bike plays the role of P and the three weather possibilities are Q_1, Q_2, and Q_3.)

Next, the Ratio Formula connects conditional credences to Kolmogorov's axioms in a special way. Conditional credence is a two-place function, taking in an ordered pair of propositions and yielding a real number. Now suppose we designate some particular proposition R as our condition, and look at all of an agent's credences conditional on R. We now have a one-place function (because the second place has been filled by R) that we can think of as a distribution over the propositions in $\mathcal{L}$. Remarkably, if the agent's unconditional credences satisfy the probability axioms, then the Ratio Formula requires this conditional distribution $cr(\cdot \mid R)$ to satisfy those axioms as well. More formally, for any proposition R in $\mathcal{L}$ such that $cr(R) > 0$, the following will all be true:

- For any proposition P in $\mathcal{L}$, $cr(P \mid R) \geq 0$.
- For any tautology T in $\mathcal{L}$, $cr(\mathsf{T} \mid R) = 1$.
- For any mutually exclusive propositions P and Q in $\mathcal{L}$,
  $cr(P \vee Q \mid R) = cr(P \mid R) + cr(Q \mid R)$.

(You'll prove these three facts in Exercise 3.4.)

Knowing that a conditional credence distribution is a probability distribution can be a handy shortcut. (It also has a significance for updating credences that we'll discuss in Chapter 4.) Because it's a probability distribution, a conditional credence distribution must satisfy all the consequences of the probability axioms we saw in Section 2.2.1. For example, if I tell you that $cr(P \mid R) = 0.7$, you can immediately tell that $cr(\sim P \mid R) = 0.3$, by the following conditional implementation of the Negation rule:

$$cr(\sim P \mid R) = 1 - cr(P \mid R) \tag{3.6}$$

Similarly, Entailment tells us that if $P \vDash Q$, then $cr(P \mid R) \leq cr(Q \mid R)$.

One special conditional distribution is worth investigating at this point: What happens when the condition R is a tautology? Imagine I ask you to report your unconditional credences in a bunch of propositions. Then I ask you to assign credences to those same propositions conditional on the further supposition of... nothing. I give you nothing more to suppose. Clearly you'll just report back to me the same credences. Bayesians represent vacuous information as a tautology, so this means that a rational agent's credences conditional on a tautology equal her unconditional credences. In other words, for any P in $\mathcal{L}$,

$$cr(P \mid \mathsf{T}) = cr(P) \tag{3.7}$$

This fact (whose proof I'll leave to the reader) will be important to our theory of updating later on.[3]

3.1.3 Bayes's Theorem

The most famous consequence of the Ratio Formula and Kolmogorov's axioms is

Bayes's Theorem: For any H and E in $\mathcal{L}$,

$$cr(H \mid E) = \frac{cr(E \mid H) \cdot cr(H)}{cr(E)}$$

The first thing to say about Bayes's Theorem is *that it is a theorem*—it can be proven straightforwardly from the axioms and Ratio Formula. This is worth remembering, because there is a great deal of controversy about how Bayesians *apply* the theorem. (The significance they attach to this theorem is why Bayesians came to be called "Bayesians".)

What philosophical significance could attach to an equation that is, in the end, just a truth of mathematics? The theorem was first articulated by the Reverend Thomas Bayes in the 1700s.[4] Prior to Bayes, much of probability theory was concerned with problems of **direct inference**. Direct inference starts with the supposition of some probabilistic hypothesis, then asks how likely that hypothesis makes a particular experimental result. You probably learned to solve many direct inference problems in school, such as "Suppose I flip a fair coin 20 times; how likely am I to get exactly 19 heads?" Here the probabilistic hypothesis H says that the coin is fair, while the experimental result E is that 20 flips yield exactly 19 heads. Your credence that the experimental result will occur on the supposition that the hypothesis is true—$cr(E\,|\,H)$—is called the **likelihood**.[5]

Yet Bayes was also interested in **inverse inference**. Instead of making suppositions about hypotheses and determining probabilities of courses of evidence, his theorem allows us to calculate probabilities of hypotheses from suppositions about evidence. Instead of calculating the likelihood $cr(E\,|\,H)$, Bayes's Theorem shows us how to calculate $cr(H\,|\,E)$. A problem of inverse inference might ask, "Suppose a coin comes up heads on exactly 19 of 20 flips; how probable is it that the coin is fair?"

Assessing the significance of Bayes's Theorem, Hans Reichenbach wrote:

> The *method of indirect evidence*, as this form of inquiry is called, consists of inferences that on closer analysis can be shown to follow the structure of the rule of Bayes. The physician's inferences, leading from the observed symptoms to the diagnosis of a specified disease, are of this type; so are the inferences of the historian determining the historical events that must be assumed for the explanation of recorded observations; and, likewise, the inferences of the detective concluding criminal actions from inconspicuous observable data.... Similarly, the general inductive inference from observational data to the validity of a given scientific theory must be regarded as an inference in terms of Bayes' rule.   (Reichenbach 1935/1949, pp. 94–5)[6]

Here's an example of inverse inference: You're a biologist studying a particular species of fish, and you want to know whether the genetic allele coding for blue fins is dominant or recessive. Based on some other work you've done on fish, you're leaning toward recessive—initially you assign a 0.4 credence that the blue-fin allele is dominant. Given some background assumptions we won't worry about here,[7] a direct inference from the theory of genetics tells you that if the allele is dominant, roughly three out of four species members will have blue fins; if the allele is recessive blue fins will appear on roughly 25% of the fish. But you're going to perform an inverse inference, making experimental observations to decide between genetic hypotheses. You will capture fish from the species at random and examine their fins. How significant will your first observation be to your credences in dominant versus recessive? When you contemplate various ways that observation might turn out, how should supposing one outcome or the other affect your credences about the allele? Before we do the calculation, try estimating how confident you should be that the allele is dominant on the supposition that the first fish you observe has blue fins.

In this example our hypothesis H will be that the blue-fin allele is dominant. The evidence E to be supposed is that a randomly drawn fish has blue fins. We want to calculate the **posterior** value $cr(H | E)$. This value is called the "posterior" because it's your credence in the hypothesis H *after* the evidence E has been supposed. In order to calculate this posterior, Bayes's Theorem requires the values of $cr(E | H)$, $cr(H)$, and $cr(E)$.

$cr(E | H)$ is the likelihood of drawing a blue-finned fish on the hypothesis that the allele is dominant. On the supposition that the allele is dominant, 75% of the fish have blue fins, so your $cr(E | H)$ value should be 0.75. The other two values are known as **priors**; they are your unconditional credences in the hypothesis and the evidence *before* anything is supposed. We already said that your prior in the blue-fin dominant hypothesis H is 0.4. So $cr(H)$ is 0.4. But what about the prior in the evidence? How confident are you before observing any fish that the first one you draw will have blue fins?

Here we can apply the Law of Total Probability to the partition containing H and $\sim H$. This yields:

$$cr(E) = cr(E | H) \cdot cr(H) + cr(E | \sim H) \cdot cr(\sim H) \qquad (3.8)$$

The values on the right-hand side are all either likelihoods, or priors related to the hypothesis. These values we can easily calculate. So

$$cr(E) = 0.75 \cdot 0.4 + 0.25 \cdot 0.6 = 0.45 \qquad (3.9)$$

Plugging all these values into Bayes's Theorem gives us

$$cr(H | E) = \frac{cr(E | H) \cdot cr(H)}{cr(E)} = \frac{0.75 \cdot 0.4}{0.45} = 2/3 \qquad (3.10)$$

Observing a single fish has the potential to change your credences substantially. On the supposition that the fish you draw has blue fins, your credence that the blue-fin allele is dominant goes from its prior value of $2/5$ to a posterior of $2/3$.

Again, all of this is pure mathematics from a set of axioms that are rarely disputed. So why has Bayes's Theorem been the focus of controversy? One issue is the role Bayesians give the theorem in *updating* attitudes over time; we'll return to that application in Chapter 4. But the main idea Bayesians take from Bayes—the idea that has proven controversial—is that probabilistic inverse inference is the key to induction. Bayesians think the primary way we ought to draw conclusions from data—how we ought to reason about scientific hypotheses, say, on the basis of experimental evidence—is by calculating posterior credences using Bayes's Theorem. This view stands in direct conflict with other statistical methods, such as frequentism and likelihoodism. Advocates of those methods also have deep concerns about where Bayesians get the priors that Bayes's Theorem requires. Once we've considerably deepened our understanding of Bayesian epistemology, we will discuss those concerns in Chapter 13, and assess frequentism and likelihoodism as alternatives to Bayesianism.

Before moving on, I'd like to highlight two useful alternative forms of Bayes's Theorem. We've just seen that calculating the prior of the evidence—$cr(E)$—can be easier if we break it up using the Law of Total Probability. Incorporating that manuever into Bayes's Theorem yields

$$cr(H \mid E) = \frac{cr(E \mid H) \cdot cr(H)}{cr(E \mid H) \cdot cr(H) + cr(E \mid \sim H) \cdot cr(\sim H)} \tag{3.11}$$

When a particular hypothesis H is under consideration, its negation $\sim H$ is known as the **catchall** hypothesis. So this form of Bayes's Theorem calculates the posterior in the hypothesis from the priors and likelihoods of the hypothesis and its catchall.

In other situations we have multiple hypotheses under consideration instead of just one. Given a finite partition of n hypotheses $\{H_1, H_2, \ldots, H_n\}$, the Law of Total Probability transforms the denominator of Bayes's Theorem to yield

$$cr(H_i \mid E) = \frac{cr(E \mid H_i) \cdot cr(H_i)}{cr(E \mid H_1) \cdot cr(H_1) + cr(E \mid H_2) \cdot cr(H_2) + \ldots + cr(E \mid H_n) \cdot cr(H_n)} \tag{3.12}$$

This version allows you to calculate the posterior of one particular hypothesis H_i in the partition from the priors and likelihoods of all the hypotheses.

3.2 Relevance and independence

Arturo doesn't believe in hocus pocus; from his point of view, information about what a clairvoyant predicts is irrelevant to determining how a coin flip will come out. So supposing that a clairvoyant predicts heads makes no difference to Arturo's confidence in a heads outcome. If C says the clairvoyant predicts heads, H says the coin lands heads, and cr_A is Arturo's credence distribution, we have

$$cr_A(H \mid C) = 1/2 = cr_A(H) \tag{3.13}$$

Generalizing this idea yields a key definition: Proposition P is **probabilistically independent** of proposition Q relative to distribution cr just in case

$$cr(P \mid Q) = cr(P) \tag{3.14}$$

In this case Bayesians also say that Q is **irrelevant** to P. When Q is irrelevant to P, supposing Q leaves an agent's credence in P unchanged.[8]

Notice that probabilistic independence is always relative to a distribution cr. The very same propositions P and Q might be independent relative to one distribution but dependent relative to another. (Relative to Arturo's credences the clairvoyant's prediction is irrelevant to the flip outcome, but relative to the credences of his friend Baxter—who believes in psychic powers—it is not.) In what follows I may omit reference to a particular distribution when context makes it clear, but you should keep the relativity of independence to a probability distribution in the back of your mind.

While Equation (3.14) will be our official *definition* of probabilistic independence, there are many equivalent tests for independence. Given the probability axioms and Ratio Formula, the following equations are all true just when Equation (3.14) is:

$$cr(P) = cr(P \mid \sim Q) \tag{3.15}$$
$$cr(P \mid Q) = cr(P \mid \sim Q) \tag{3.16}$$
$$cr(Q \mid P) = cr(Q) = cr(Q \mid \sim P) \tag{3.17}$$
$$cr(P \,\&\, Q) = cr(P) \cdot cr(Q) \tag{3.18}$$

The equivalence of Equations (3.14) and (3.15) tells us that Q is probabilistically independent of P just in case $\sim Q$ is. The equivalence of (3.14) and (3.17) shows us that independence is symmetric: if supposing Q makes no difference to an agent's credence in P, then supposing P won't change that agent's attitude toward Q. Finally, Equation (3.18) embodies a useful probability rule:

Multiplication: P and Q are probabilistically independent relative to cr if and only if $cr(P \,\&\, Q) = cr(P) \cdot cr(Q)$.

(Some authors define probabilistic independence using this biconditional, but we will define independence using Equation (3.14), then treat Multiplication as a consequence.)

We can calculate the probability of a conjunction by multiplying the probabilities of its conjuncts only when those conjuncts are *independent*. This trick will not work for any arbitrary propositions. The general formula for probability in a conjunction can be derived quickly from the Ratio Formula:

$$cr(P \,\&\, Q) = cr(P \,|\, Q) \cdot cr(Q) \qquad (3.19)$$

When P and Q are probabilistically independent, the first term on the right-hand side equals $cr(P)$.

It's important not to get Multiplication and Finite Additivity confused. Finite Additivity says that the credence of a *disjunction* is the *sum* of the credences of its *mutually exclusive* disjuncts. Multiplication says that the credence of a *conjunction* is the *product* of the credences of its *independent* conjuncts. If I flip a fair coin twice in succession, heads on the first flip and heads on the second flip are independent, while heads on the first flip and tails on the first flip are mutually exclusive.

When two propositions fail to be probabilistically independent (relative to a particular distribution), we say those propositions are **relevant** to each other. Replace the "=" signs in Equations (3.14) through (3.18) with ">" signs and you have tests for Q's being **positively relevant** to P. Once more the tests are equivalent—if any of the resulting inequalities is true, all of them are. Q is positively relevant to P when assuming Q makes you more confident in P. For example, since Baxter believes in mysticism, he takes the clairvoyant's predictions to be highly relevant to the outcome of the coin flip—supposing that the clairvoyant has predicted heads takes him from equanimity to near-certainty in a heads outcome. Baxter assigns

$$cr_B(H \,|\, C) = 99/100 > 1/2 = cr_B(H) \qquad (3.20)$$

Like independence, positive relevance is symmetric. Supposing that the coin came up heads will make Baxter highly confident that the clairvoyant predicted it would.

Similarly, replacing the "=" signs with "<" signs above yields tests for **negative relevance**. For Baxter, the clairvoyant's predicting heads is negatively relevant to the coin's coming up tails. Like positive correlation, negative correlation is symmetric (supposing a tails outcome makes Baxter less confident in a heads prediction). Note also that there are many synonyms in the statistics community for "relevance". Instead of finding "positively/ negatively relevant" terminology, you'll sometimes find "positively/negatively dependent", "positively/negatively correlated", or even "correlated/ anti-correlated".

The strongest forms of positive and negative relevance are entailment and refutation. Suppose a hypothesis H has nonextreme prior credence. If a particular piece of evidence E *entails* the hypothesis, the probability axioms and Ratio Formula tell us that

$$cr(H \mid E) = 1 \qquad\qquad (3.21)$$

Supposing E takes H from a middling credence to the highest credence allowed. Similarly, if E refutes H (what philosophers of science call **falsification**), then

$$cr(H \mid E) = 0 \qquad\qquad (3.22)$$

Relevance will be most important to us because of its connection to confirmation, the Bayesian notion of evidential support. A piece of evidence confirms a hypothesis only if it's relevant to that hypothesis. Put another way, learning a piece of evidence changes a rational agent's credence in a hypothesis only if that evidence is relevant to the hypothesis. (Much more on this later.)

3.2.1 Conditional independence and screening off

The definition of probabilistic independence compares an agent's conditional credence in a proposition to her unconditional credence in that proposition. But we can also compare conditional credences. When Baxter, who believes in the occult, hears a clairvoyant's prediction about the outcome of a fair coin flip, he takes it to be highly correlated with the true flip outcome. But what if we ask Baxter to suppose that this particular clairvoyant is an impostor? Once he supposes the clairvoyant is an impostor, Baxter may take the clairvoyant's predictions to be completely irrelevant to the flip outcome. Let C be the proposition that the clairvoyant predicts heads, H be the proposition that the coin comes up heads, and I be the proposition that the clairvoyant is an impostor. It's possible for Baxter's credences to satisfy both of the following equations at once:

$$\text{cr}(H \mid C) > \text{cr}(H) \tag{3.23}$$

$$\text{cr}(H \mid C \,\&\, I) = \text{cr}(H \mid I) \tag{3.24}$$

Inequality (3.23) tells us that unconditionally, Baxter takes C to be relevant to H. But conditional on the supposition of I, C becomes independent of H (Equation (3.24)); once Baxter has supposed I, adding C to his suppositions doesn't affect his confidence in H.

Generalizing this idea yields the following definition of **conditional independence**: P is probabilistically independent of Q conditional on R just in case

$$\text{cr}(P \mid Q \,\&\, R) = \text{cr}(P \mid R) \tag{3.25}$$

If this equality fails to hold, we say that Q is relevant to P conditional on R.

One more piece of terminology: We will say that R **screens off** P from Q when Q is unconditionally relevant to P, but irrelevant to P conditional on each of R and $\sim R$. That is, R screens off P from Q just in case all three of the following are satisfied:

$$\text{cr}(P \mid Q) \neq \text{cr}(P) \tag{3.26}$$

$$\text{cr}(P \mid Q \,\&\, R) = \text{cr}(P \mid R) \tag{3.27}$$

$$\text{cr}(P \mid Q \,\&\, {\sim}R) = \text{cr}(P \mid {\sim}R) \tag{3.28}$$

When these equations are met, P and Q are correlated but supposing either R or $\sim R$ makes that correlation disappear.[9]

Conditional independence and screening off are both best understood through real-world examples. We'll see a number of those in the next few sections.

3.2.2 The Gambler's Fallacy

People often act as if future chancy events will "compensate" for unexpected past results. When a good hitter strikes out many times in a row, someone will say he's "due" for a hit. If a fair coin comes up heads nineteen times in a row, many people become more confident that the next outcome will be tails.

This mistake is known as the **Gambler's Fallacy**.[10] A person who makes the mistake is thinking along something like the following lines: In twenty flips of a fair coin, it's more probable to get nineteen heads and one tail than it is to get twenty heads. So having seen nineteen heads, it's much more likely that the twentieth flip will come up tails.

This person is providing the right answer to the wrong question. If the question is "When a fair coin is flipped twenty times, is it more likely that you'll get a *total* of nineteen heads and one tail than it is that you'll get twenty heads?", the answer to that question is "yes"—in fact, it's twenty times as likely! But that's the wrong question to ask in this case. Instead of wondering what sorts of total outcomes are probable when one flips a fair coin twenty times, in this case it's more appropriate to ask: *given* that the coin has already come up heads nineteen times, how confident are we that the twentieth flip will be tails? This is a question about our conditional credence

$$\text{cr(20th flip tails} \mid \text{first 19 flips heads)} \tag{3.29}$$

How should we calculate this conditional credence? Ironically, it might be more reasonable to make a mistake in the *opposite* direction from the Gambler's Fallacy. If I see a coin come up heads nineteen times, shouldn't that make me suspect that it's biased toward heads? If anything, shouldn't supposing nineteen consecutive heads make me *less* confident that the twentieth flip will come up tails?

This line of reasoning would be appropriate to the present case if we hadn't stipulated that the coin is fair. For a rational agent, the outcome of the twentieth flip is probabilistically independent of the outcomes of the first nineteen flips conditional on the fact that the coin is fair. That is,

$$\begin{aligned} \text{cr(20th flip tails} \mid \text{first 19 flips heads \& fair coin)} = \\ \text{cr(20th flip tails} \mid \text{fair coin)} \end{aligned} \tag{3.30}$$

We can justify this equation as follows: Typically, information that a coin came up heads nineteen times in a row would alter your opinion about whether it's a fair coin. Changing your opinion about whether it's a fair coin would then affect your prediction for the twentieth flip. So typically, information about the first nineteen flips alters your credences about the twentieth flip *by way of* your opinion about whether the coin is fair. But if you've already established that the coin is fair, information about the first nineteen flips has no further significance for your prediction about the twentieth. So conditional on the coin's being fair, the first nineteen flips' outcomes are irrelevant to the outcome of the twentieth flip.

The left-hand side of Equation (3.30) captures the correct question to ask about the Gambler's Fallacy case. The right-hand side is easy to calculate; it's 1/2. So after seeing a coin known to be fair come up heads nineteen times, we should be 1/2 confident that the twentieth flip will be tails.[11]

3.2.3 Probabilities are weird! Simpson's Paradox

Perhaps you're too much of a probabilistic sophisticate to ever commit the Gambler's Fallacy. Perhaps you successfully navigated Tversky and Kahneman's Conjunction Fallacy (Section 2.2.4) as well. But even probability experts sometimes have trouble with the counterintuitive relations that arise between conditional and unconditional dependence.

Here's an example of how odd things can get: In a famous case, the University of California, Berkeley's graduate departments were investigated for gender bias in admissions. The concern arose because in 1973 about 44% of overall male applicants were admitted to graduate school at Berkeley, while only 35% of female applicants were. Yet when the graduate departments (where admissions decisions are made) were studied one at a time, it turned out that individual departments either were admitting men and women at roughly equal rates, or in some cases were admitting a higher percentage of female applicants.

This is an example of **Simpson's Paradox**, in which probabilistic dependencies point in one direction conditional on each member of a partition, yet point the opposite way unconditionally. A Simpson's Paradox case involves a collection with a number of subgroups. Each of the subgroups displays the same correlation between two traits. Yet when we examine the collection as a whole, that correlation is reversed![12]

To see how this can happen, consider another example: Over the course of the 2016–2017 NBA season, Houston Rockets player James Harden made a higher percentage of his two-point shots than Toronto Raptors player DeMar DeRozan did. Harden also made a higher percentage of his three-point shots than DeRozan. Yet when we look at all the shots attempted (both two- and three-pointers), DeRozan made a higher percentage than Harden overall.[13]

Here are the data for the two players:

	Two-pointers		Three-pointers		Combined	
DeRozan	688/1421	48.4%	33/124	26.6%	721/1545	46.7%
Harden	412/777	53.0%	262/756	34.7%	674/1533	44.0%

The second number in each box is the number of attempts; the first is the number of makes; the third is the percentage (makes divided by attempts). Looking at the table, you can see how Harden managed to shoot better than DeRozan from each distance yet have a worse shooting percentage overall. Since three-pointers are more difficult to make than two-pointers, each player made a much higher percentage of his two-point attempts than his three-point attempts. DeRozan shot over 90% of his shots from the easier-to-make two-point distance. Harden, on the other hand, shot almost half of his shots from downtown. Harden was taking many more low-percentage shots than DeRozan, so even though Harden was better at those shots, his overall percentage suffered.

Scrutiny revealed a similar effect in Berkeley's 1973 admissions data. Bickel, Hammel, and O'Connell (1975) concluded, "The proportion of women applicants tends to be high in departments that are hard to get into and low in those that are easy to get into." Although individual departments were admitting women at comparable rates to men, female applications were less successful overall because more were directed at departments with low admission rates.[14]

Simpson's Paradox can be thought of entirely in terms of numerical proportions, as we've just done with the basketball and admissions examples. But these examples can also be analyzed using conditional probabilities. Suppose, for instance, that you are going to select a Harden or DeRozan shot attempt at random from the 2016–2017 season, making your selection so that each of the 3,078 attempts they put up together is equally likely to be selected. How confident should you be that the selected attempt will be a make? How should that confidence change on the supposition that a DeRozan attempt is selected, or a two-point attempt?

Below is a probability table for your credences, assembled from the real-life statistics above. Here D says that it's a DeRozan attempt; 2 says it's a two-pointer; and M says it's a make. (Given the pool from which we're sampling, $\sim D$ means a Harden attempt and ~ 2 means it's a three-pointer.)

D	2	M	cr
T	T	T	688/3078
T	T	F	733/3078
T	F	T	33/3078
T	F	F	91/3078
F	T	T	412/3078
F	T	F	365/3078
F	F	T	262/3078
F	F	F	494/3078

A bit of calculation with this probability table reveals the following:

$$\text{cr}(M \mid D) > \text{cr}(M \mid {\sim}D) \tag{3.31}$$

$$\text{cr}(M \mid D \,\&\, 2) < \text{cr}(M \mid {\sim}D \,\&\, 2) \tag{3.32}$$

$$\text{cr}(M \mid D \,\&\, {\sim}2) < \text{cr}(M \mid {\sim}D \,\&\, {\sim}2) \tag{3.33}$$

If you're selecting an attempt from the total sample, DeRozan is more likely to make it than Harden. Put another way, DeRozan's taking the attempt is unconditionally positively relevant to its being made (Equation (3.31)). But DeRozan's shooting is negatively relevant to a make conditional on each of the two distances (Equations (3.32) and (3.33)). If you're selecting from only the two-pointers, or from only the three-pointers, the shot is more likely to be made if it's attempted by Harden.

3.2.4 Correlation and causation

You may have heard the expression "correlation is not causation." People typically use this expression to point out that just because two events have both occurred—and maybe occurred in close spatio-temporal proximity—that doesn't mean they had anything to do with each other. But "correlation" is a technical term in probability discussions. The propositions describing two events may both be true, or you might have high credence in both of them, yet they still might not be probabilistically correlated. For the propositions to be correlated, supposing one to be true must *change* the probability of the other. I'm confident that I'm under 6 feet tall and that my eyes are blue, but that doesn't mean I take those facts to be correlated.

Once we've understood probabilistic correlation correctly, does its presence always indicate a causal connection? When two propositions about empirical events are correlated, must the event described by one cause the event described by the other? Hans Reichenbach offered a classic counterexample to this proposal. He wrote:

> Suppose two geysers which are not far apart spout irregularly, but throw up their columns of water always at the same time. The existence of a subterranean connection of the two geysers with a common reservoir of hot water is then practically certain.   (1956, p. 158)

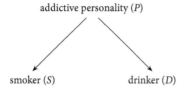

Figure 3.2 A causal fork

If you've noticed that two nearby geysers always spout simultaneously, seeing one spout will increase your confidence that the other is spouting as well. So your credences about the geysers are correlated. But you don't think one geyser's spouting *causes* the other to spout. Instead, you hypothesize an unobserved reservoir of hot water that is the **common cause** of both spouts.

Reichenbach proposed a famous principle about empirically correlated events:

Principle of the Common Cause: When event outcomes are probabilistically correlated, either one causes the other or they have a common cause.[15]

In addition to this principle, he offered a key mathematical insight about causation: a common cause screens its effects off from each other.

Let's work through an example of this insight concerning causation and screening off. Suppose the proposition that a particular individual is a drinker is positively relevant to the proposition that she's a smoker. According to the Principle of the Common Cause, there must be some causal link between these propositions. Perhaps drinking causes smoking—drinking creates situations in which one is more likely to smoke—or vice versa. Or they may be linked through a common cause: maybe being a smoker and being a drinker are both promoted by an addictive personality, which we can imagine results from a genetic endowment unaffected by one's behavior. (See Figure 3.2; the arrows indicate causal influence.)

Imagine the latter explanation is true, and moreover is the *only* true explanation of the correlation between drinking and smoking. That is, being a smoker and being a drinker are positively correlated only due to their both being caused by an addictive personality. Given this assumption, let's take a particular subject whose personality you're unsure about, and consider what happens to your credences when you make various suppositions about her.

If you begin by supposing that the subject drinks, this will make you more confident that she smokes—*but only because it makes you more confident*

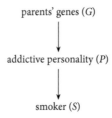

parents' genes (*G*)

addictive personality (*P*)

smoker (*S*)

Figure 3.3 A causal chain

that the subject has an addictive personality. On the other hand, you might start by supposing that the subject has an addictive personality. That will certainly make you more confident that she's a smoker. But once you've made that adjustment, going on to suppose that she's a drinker won't affect your confidence in smoking. Information about drinking affects your smoking opinions only *by way of* helping you figure out whether she has an addictive personality, and the answer to the personality question was filled in by your initial supposition. Once an addictive personality is supposed, drinking has no further relevance to smoking. (Compare: Once a coin is supposed to be fair, the outcomes of its first nineteen flips have no relevance to the outcome of the twentieth.) Drinking becomes probabilistically independent of smoking conditional on suppositions about whether the subject has an addictive personality. That is,

$$cr(S \mid D) > cr(S) \tag{3.34}$$

$$cr(S \mid D \,\&\, P) = cr(S \mid P) \tag{3.35}$$

$$cr(S \mid D \,\&\, {\sim}P) = cr(S \mid {\sim}P) \tag{3.36}$$

Causal forks (as in Figure 3.2) give rise to screening off. *P* is a common cause of *S* and *D*, so *P* screens off *S* from *D*.

But that's not the only way screening off can occur. Consider Figure 3.3. Here we've focused on a different portion of the causal structure. Imagine that the subject's parents' genes causally influence whether she has an addictive personality, which in turn causally promotes smoking. Now her parents' genetics are probabilistically relevant to the subject's smoking, but that correlation is screened off by facts about her personality. Again, if you're uncertain whether the subject's personality is addictive, facts about her parents' genes will affect your opinion of whether she's a smoker. But once you've made a firm supposition about the subject's personality, suppositions about her parents' genetics have no further influence on your smoking opinions. In equation form:

$$\text{cr}(S \mid G) > \text{cr}(S) \tag{3.37}$$
$$\text{cr}(S \mid G \,\&\, P) = \text{cr}(S \mid P) \tag{3.38}$$
$$\text{cr}(S \mid G \,\&\, {\sim}P) = \text{cr}(S \mid {\sim}P) \tag{3.39}$$

P screens off S from G.[16]

Relevance, conditional relevance, and causation can interact in very complex ways.[17] My goal here has been to introduce the main ideas and terminology employed in their analysis. The state of the art in this field has come a long way since Reichenbach; computational tools now available can look at statistical correlations among a large number of variables and hypothesize a causal structure lying beneath them. The resulting causal diagrams are known as **Bayes Nets**, and have practical applications from satellites to health care to car insurance to college admissions.

These causal methods all start from Reichenbach's insight that common causes screen off their effects. And what of his more metaphysically radical Principle of the Common Cause? It remains highly controversial.

3.3 Conditional credences and conditionals

I now want to circle back and get clearer on the nature of conditional credence. First, it's important to note that the conditional credences we've been discussing are indicative, not subjunctive. This distinction is familiar from the theory of conditional propositions. Compare:

> If Shakespeare didn't write *Hamlet*, then someone else did.
> If Shakespeare hadn't written *Hamlet*, then someone else would have.

The former conditional is indicative, while the latter is subjunctive. The traditional distinction between these two types of conditional begins with the assumption that a conditional is evaluated by considering possible worlds in which the antecedent is satisfied, then checking whether the consequent is true in those worlds as well. In evaluating an indicative conditional, the antecedent worlds are restricted to those among the agent's doxastic possibilities.[18] Evaluating a subjunctive conditional, on the other hand, permits *counterfactual* reasoning involving antecedent worlds considered non-actual. So when you assess the subjunctive conditional above, you are allowed to consider worlds that make the antecedent true by making *Hamlet* never exist. But when evaluating the indicative conditional, you have to take into account that *Hamlet* actually does exist, and entertain only worlds in which that's true. So you consider bizarre "author-conspiracy" worlds that, while far-fetched, satisfy the antecedent and are among your current doxastic possibilities. In the end, I'm guessing you take the indicative conditional to be true but the subjunctive to be false.

Now suppose I ask for your credence in the proposition that someone wrote *Hamlet*, conditional on the supposition that Shakespeare didn't. This value will be high, again because you take *Hamlet* to exist. In assigning this conditional credence, you aren't bringing into consideration possible worlds you'd otherwise ruled out (such as *Hamlet*-free worlds). Instead, you're focusing in on the narrow set of author-conspiracy worlds you currently entertain. As we saw in Figure 3.1, assigning a conditional credence strictly narrows the worlds under consideration; it doesn't expand your attention to worlds previously ruled out. Thus the conditional credences discussed in this book—and typically discussed in the Bayesian literature—are indicative rather than subjunctive.[19]

Are there more features of conditional propositions that can help us understand conditional credences? Might we understand conditional credences *in terms of* conditionals? Initiating his study of conditional degrees of belief, F.P. Ramsey warned against assimilating them to conditional propositions:

> We are also able to define a very useful new idea—"the degree of belief in p given q". This does not mean the degree of belief in "If p then q", or that in "p entails q", or that which the subject would have in p if he knew q, or that which he ought to have.   (1931, p. 82)[20]

Yet many authors failed to heed Ramsey's warning. It's very tempting to equate conditional credences with some simple combination of conditional propositions and unconditional credences. For example, when I ask, "How confident are you in P given Q?", it's easy to hear that as "Given Q, how confident are you in P?" or just "If Q is true, how confident are you in P?" This simple slide might suggest that for any real number r and propositions P and Q,

$$\text{``}cr(P \mid Q) = r\text{''} \text{ is equivalent to } \text{``}Q \to cr(P) = r\text{''} \qquad (3.40)$$

Here I'm using the symbol "$\to$" to represent some kind of conditional. For the reasons discussed above, it should be an indicative conditional. But it need not be the material conditional symbolized by "$\supset$"; many authors think the material conditional's truth-function fails to accurately represent the meaning of natural-language indicative conditionals.

Endorsing the equivalence in (3.40) would require serious changes to the traditional logic of conditionals. We can demonstrate this in two ways. First, we usually take indicative conditionals to satisfy the disjunctive syllogism rule. (The material conditional certainly does!) This rule tells us that

$$\text{"}X \to Z\text{" and "}Y \to Z\text{" jointly entail "}(X \lor Y) \to Z\text{"} \qquad (3.41)$$

for any propositions X, Y, and Z. Thus for any propositions A, B, and C and constant k we have

$$\text{"}A \to [cr(C) = k]\text{" and "}B \to [cr(C) = k]\text{" entail "}(A \lor B) \to [cr(C) = k]\text{"} \qquad (3.42)$$

Combining (3.40) and (3.42) yields

$$\text{"}cr(C\,|\,A) = k\text{" and "}cr(C\,|\,B) = k\text{" entail "}cr(C\,|\,A \lor B) = k\text{"} \qquad (3.43)$$

(3.43) may look appealing, as a sort of probabilistic analog of disjunctive syllogism. But it's false. Not only can one design a credence distribution satisfying the probability axioms and Ratio Formula such that $cr(C\,|\,A) = k$ and $cr(C\,|\,B) = k$ but $cr(C\,|\,A \lor B) \neq k$; one can even describe real-life examples in which it's rational for an agent to assign such a distribution. (See Exercise 3.14.) The failure of (3.43) is another case in which credences confound expectations developed by our experiences with classificatory states.

Second, we usually take indicative conditionals to satisfy *modus tollens*. But consider the following facts about me: Unconditionally, I am highly confident that I will be alive tomorrow. But conditional on the proposition that the sun just exploded, my confidence that I will be alive tomorrow is very low. Given these facts, *modus tollens*, and (3.40), I could run the following argument:

cr(alive tomorrow | sun exploded) is low.	[given]	(3.44)
If the sun exploded, cr(alive tomorrow) is low.	[(3.44), (3.40)]	(3.45)
cr(alive tomorrow) is high.	[given]	(3.46)
The sun did not explode.	[*modus tollens*]	(3.47)

While intriguing for its promise of astronomy by introspection, this argument is unsound. So I conclude that as long as indicative conditionals satisfy classical logical rules such as disjunctive syllogism and *modus tollens*, any analysis of conditional credences in terms of conditionals that uses (3.40) must be false.[21]

Perhaps we've mangled the transition from conditional credences to conditional propositions. Perhaps we should hear "How confident are you in P given Q?" as "How confident are you in 'P, given Q'?", which is in turn "How confident are you in 'If Q, then P'?" Maybe a conditional credence is a credence in a conditional. Or perhaps more weakly: an agent assigns a particular conditional credence value whenever she unconditionally assigns that value to a conditional. In symbols, the proposal is that

$$\text{"}cr(P \mid Q) = r\text{" is equivalent to "}cr(Q \to P) = r\text{"} \tag{3.48}$$

for any real r, any propositions P and Q, any credence distribution cr, and some indicative conditional $\to$. If true, this equivalence would offer another possibility for analyzing conditional credences in terms of unconditional credences and conditionals.

We can quickly show that (3.48) fails if "$\to$" is read as the material conditional $\supset$. Under the material reading, the proposal entails that

$$cr(P \mid Q) = cr(Q \supset P) \tag{3.49}$$

Using the probability calculus and Ratio Formula, we can show that Equation (3.49) holds only when $cr(Q) = 1$ or $cr(Q \supset P) = 1$. (See Exercise 3.15.) This is a *triviality result*: It shows that Equation (3.49) can hold only in trivial cases, namely over the narrow range of conditionals for which the agent is either certain of the antecedent or certain of the conditional itself. Equation (3.49) does not express a truth that holds for *all* conditional credences in *all* propositions; nor does (3.48) when "$\to$" is read materially.

Perhaps the equivalence in (3.48) can be saved from this objection by construing its "$\to$" as something other than a material conditional. But David Lewis (1976) provided a clever objection that works whichever conditional $\to$ we choose. Begin by selecting arbitrary propositions P and Q. We then derive the following from the proposal in Equation (3.48):

$$cr(Q \to P) = cr(P \mid Q) \qquad\qquad \text{[from (3.48)]} \qquad (3.50)$$

$$cr(Q \to P \mid P) = cr(P \mid Q \,\&\, P) \qquad\qquad \text{[see below]} \qquad (3.51)$$

$$cr(Q \to P \mid P) = 1 \qquad\qquad \text{[}Q \,\&\, P \text{ entails } P\text{]} \qquad (3.52)$$

$$cr(Q \to P \mid {\sim}P) = cr(P \mid Q \,\&\, {\sim}P) \qquad\qquad \text{[see below]} \qquad (3.53)$$

$$cr(Q \to P \mid {\sim}P) = 0 \qquad\qquad \text{[}Q \,\&\, {\sim}P \text{ refutes } P\text{]} \qquad (3.54)$$

$$cr(Q \to P) = cr(Q \to P \,|\, P) \cdot cr(P) +$$
$$cr(Q \to P \,|\, {\sim}P) \cdot cr({\sim}P) \qquad \text{[Law of Total Prob.]} \qquad (3.55)$$
$$cr(Q \to P) = 1 \cdot cr(P) + 0 \cdot cr({\sim}P) \qquad \text{[(3.52), (3.54), (3.55)]} \qquad (3.56)$$
$$cr(Q \to P) = cr(P) \qquad\qquad\qquad\qquad\qquad\qquad (3.57)$$
$$cr(P \,|\, Q) = cr(P) \qquad\qquad\qquad\qquad \text{[(3.50)]} \qquad (3.58)$$

Some of these lines require explanation. The idea of lines (3.51) and (3.53) is this: We've already seen that a credence distribution conditional on a particular proposition satisfies the probability axioms. This suggests that we should think of a distribution conditional on a proposition as being just like any other credence distribution. (We'll see more reason to think this in Chapter 4, note 3.) So a distribution conditional on a proposition should satisfy the proposal of (3.48) as well. If you conditionally suppose X, then under that supposition you should assign $Y \to Z$ the same credence you would assign Z were you to *further* suppose Y. In other words, anyone who maintains (3.48) should also maintain that for any X, Y, and Z,

$$cr(Y \to Z \,|\, X) = cr(Z \,|\, Y \,\&\, X) \qquad (3.59)$$

In line (3.51) the roles of X, Y, and Z are played by P, Q, and P; in line (3.53) it's ${\sim}P$, Q, and P.

Lewis has given us another triviality result. Assuming the probability axioms and Ratio Formula, the proposal in (3.48) can hold only for propositions P and Q such that $cr(P \,|\, Q) = cr(P)$. In other words, it can hold only for propositions the agent takes to be independent.[22] Or (taking things from the other end), the proposed equivalence can hold for all the conditionals in an agent's language only if the agent treats every pair of propositions in $\mathcal{L}$ as independent![23]

So a rational agent's conditional credence will not in general equal her unconditional credence in a conditional. This is not to say that conditional credences have *nothing* to do with conditionals. A popular idea now usually called "Adams's Thesis" (Adams 1965) holds that an indicative conditional $Q \to P$ is *acceptable* to a degree equal to $cr(P \,|\, Q)$.[24] But we cannot maintain that an agent's conditional credence is always equal to her credence that some conditional is *true*.

This brings us back to a proposal I discussed in Chapter 1. One might try to relate degrees of belief to binary beliefs by suggesting that whenever an agent has an r-valued credence, she has a binary belief in a traditional proposition with r as part of its content. Working out this proposal for conditional credences reveals how hopeless it is. Suppose an agent assigns $cr(P \mid Q) = r$. Would we suggest that the agent believes that if Q, then the probability of P is r? This proposal mangles the logic of conditional credences. Perhaps the agent believes that the probability of "if Q, then P" is r? Lewis's argument dooms this idea.

I said in Chapter 1 that the numerical value of an unconditional degree of belief is an attribute of the *attitude taken* toward a proposition, not a *constituent of* that proposition itself. As for conditional credences, $cr(P \mid Q) = r$ does not say that an agent takes some attitude toward a conditional proposition with a probability value in its consequent. Nor does it say that the agent takes some attitude toward a single, conditional proposition composed of P and Q. $cr(P \mid Q) = r$ says that the agent takes an r-valued attitude toward an *ordered pair* of propositions—neither of which need involve the number r.

3.4 Exercises

Unless otherwise noted, you should assume when completing these exercises that the credence distributions under discussion satisfy the probability axioms and Ratio Formula. You may also assume that whenever a conditional credence expression occurs, the condition has a nonzero unconditional credence so that the conditional credence is well defined.

Problem 3.1. 𝄐 A family has two children of different ages. Assume that each child has a probability of 1/2 of being a girl, and that the probability that the elder is a girl is independent of the probability that the younger is.

    (a) Conditional on the older child's being a girl, what's the probability that the younger one is?

    (b) Conditional on at least one child's being a girl, what's the probability that they both are?

Problem 3.2. 𝄐𝄐 Flip and Flop are playing a game. They have a fair coin that they are going to keep flipping until one of two things happens: either the coin comes up heads twice in a row, or it comes up tails followed by heads. The first time one of these things happens, the game ends—if it ended with HH, Flip wins; if it ended with TH, Flop wins.

    (a) What's the probability that Flip wins after the first two tosses of the coin? What's the probability that Flop wins after the first two tosses of the coin?

(b) Flip and Flop play their game until it ends (at which point one of them wins). What's the probability that Flop is the winner?[25]

Problem 3.3. 🎵🎵 One might think that real humans only ever assign credences that are rational numbers—and perhaps only rational numbers involving relatively small whole-number numerators and denominators. But we can construct simple conditions that *require* an irrational-valued credence distribution. For example, consider the scenario below.

You have a biased coin that you are going to flip twice in a row. Suppose your credence distribution satisfies all of the following conditions:

(i) You are equally confident that the first flip will come up heads and that the second flip will come up heads.
(ii) You treat the outcomes of the two flips as probabilistically independent.
(iii) Given what you know about the bias, your confidence that the two flips will *both* come up heads equals your confidence in all of the other outcomes put together.

Assuming your credence distribution satisfies the three conditions above, how confident are you that the first flip will come up heads?[26]

Problem 3.4. 🎵🎵 Prove that credences conditional on a particular proposition form a probability distribution. That is, prove that for any proposition R in $\mathcal{L}$ such that $cr(R) > 0$, the following three conditions hold:

(a) For any proposition P in $\mathcal{L}$, $cr(P \,|\, R) \geq 0$.
(b) For any tautology T in $\mathcal{L}$, $cr(T \,|\, R) = 1$.
(c) For any mutually exclusive propositions P and Q in $\mathcal{L}$,
$cr(P \vee Q \,|\, R) = cr(P \,|\, R) + cr(Q \,|\, R)$.

Problem 3.5. 🎵 Pink gumballs always make my sister sick. (They remind her of Pepto Bismol.) Blue gumballs make her sick half of the time (they just look unnatural), while white gumballs make her sick only one-tenth of the time. Yesterday, my sister bought a single gumball randomly selected from a machine that's one-third pink gumballs, one-third blue, and one-third white. Applying the version of Bayes's Theorem in Equation (3.12), how confident should I be that my sister's gumball was pink conditional on the supposition that it made her sick?

Problem 3.6. 🎵🎵

(a) Prove Bayes's Theorem from the probability axioms and Ratio Formula. (Hint: Start by using the Ratio Formula to write down expressions involving $cr(H \,\&\, E)$ and $cr(E \,\&\, H)$.)

(b) Exactly which unconditional credences must we assume to be positive in order for your proof to go through?

(c) Where exactly does your proof rely on the probability axioms (and not just the Ratio Formula)?

Problem 3.7. 🌙 Once more, consider the probabilistic credence distribution specified by this probability table (from Exercise 2.9):

P	Q	R	cr
T	T	T	0.1
T	T	F	0.2
T	F	T	0
T	F	F	0.3
F	T	T	0.1
F	T	F	0.2
F	F	T	0
F	F	F	0.1

Answer the following questions about this distribution:

(a) What is cr($P \mid Q$)?

(b) Relative to this distribution, is Q positively relevant to P, negatively relevant to P, or probabilistically independent of P?

(c) What is cr($P \mid R$)?

(d) What is cr($P \mid Q \& R$)?

(e) Conditional on R, is Q positively relevant to P, negatively relevant to P, or probabilistically independent of P?

(f) Does R screen off P from Q? Explain why or why not.

Problem 3.8. 🌙🌙 Prove that all the alternative statements of probabilistic independence in Equations (3.15) through (3.18) follow from our original independence definition. That is, prove that each equation (3.15) through (3.18) follows from Equation (3.14), the probability axioms, and the Ratio Formula. (Hint: Once you prove that a particular equation follows from Equation (3.14), you may use it in subsequent proofs.)

Problem 3.9. 🌙🌙 Show that probabilistic independence is not transitive. That is, provide a single probability distribution on which all of the following are true: X is independent of Y, and Y is independent of Z, but X is not independent of Z. Show that your distribution satisfies all three conditions. (For an added chili pepper of difficulty, have your distribution assign every state-description a nonzero unconditional credence.)

Problem 3.10. 🎵🎵 In the text we discussed what makes a *pair* of propositions probabilistically independent. If we have a larger collection of propositions, what does it take to make them all independent of each other? You might think all that's necessary is *pairwise independence*—for each pair within the set of propositions to be independent. But pairwise independence doesn't guarantee that each proposition will be independent of *combinations* of the others.

To demonstrate this fact, describe a real-world example (spelling out the propositions represented by X, Y, and Z) in which it would be rational for an agent to assign credences meeting all four of the following conditions:

   (i) $cr(X \mid Y) = cr(X)$
   (ii) $cr(X \mid Z) = cr(X)$
   (iii) $cr(Y \mid Z) = cr(Y)$
   (iv) $cr(X \mid Y \& Z) \neq cr(X)$

Show that your example satisfies all four conditions.

Problem 3.11. 🎵🎵 Using the program PrSAT referenced in the Further Readings for Chapter 2, find a probability distribution satisfying all the conditions in Exercise 3.10, plus the following *additional* condition: Every state-description expressible in terms of X, Y, and Z must have a nonzero unconditional cr-value.

Problem 3.12. 🎵🎵
   (a) The 2016–2017 NBA season has just ended, and you're standing on a basketball court. Suddenly aliens appear, point to a spot on the court, and say that unless someone makes the next shot attempted from that spot, they will end your life. You're highly interested in self-preservation but terrible at basketball; luckily James Harden and DeMar DeRozan are standing right there. DeRozan says, "I had a better overall shooting percentage than Harden this year, so I should attempt the shot." Given the statistics on page 70, explain why DeRozan's argument is unconvincing.
   (b) Suppose the aliens pointed to a spot that would yield a three-point attempt. You're about to hand the ball to Harden, when DeRozan says,

"I know that's a three-pointer, and Harden shot better from three-point range in general than I did this year. But from *that particular spot,* I had a better percentage than him." Is DeRozan's claim consistent with the statistics on page 70? (That is, could DeRozan's claim be true while those statistics are also accurate?) Explain why or why not.

Problem 3.13. After laying down probabilistic conditions for a causal fork, Reichenbach demonstrated that a causal fork induces correlation. Consider the following four conditions:

   (i) $cr(A \mid C) > cr(A \mid \sim C)$
   (ii) $cr(B \mid C) > cr(B \mid \sim C)$
   (iii) $cr(A \& B \mid C) = cr(A \mid C) \cdot cr(B \mid C)$
   (iv) $cr(A \& B \mid \sim C) = cr(A \mid \sim C) \cdot cr(B \mid \sim C)$

   (a) 𝄢 Assuming C is the common cause of A and B, explain what each of the four conditions means in terms of relevance, independence, conditional relevance, or conditional independence.

   (b) 𝄢𝄢 Prove that if all four conditions hold, then $cr(A \& B) > cr(A) \cdot cr(B)$.

Problem 3.14. 𝄢 In Section 3.3 I pointed out that the following statement (labeled Equation (3.43) there) does not hold for every constant k and propositions A, B, and C:

$$\text{``}cr(C \mid A) = k\text{'' and ``}cr(C \mid B) = k\text{'' entail ``}cr(C \mid A \vee B) = k\text{''}$$

   (a) Describe a real-world example (involving dice, or cards, or something more interesting) in which it's rational for an agent to assign $cr(C \mid A) = k$ and $cr(C \mid B) = k$ but $cr(C \mid A \vee B) \neq k$. Show that your example meets this description.

   (b) Prove that if A and B are mutually exclusive, then whenever $cr(C \mid A) = k$ and $cr(C \mid B) = k$ it's also the case that $cr(C \mid A \vee B) = k$.

Problem 3.15. 𝄢 Fact: For any propositions P and Q, if $cr(Q) > 0$ then $cr(Q \supset P) \geq cr(P \mid Q)$.

   (a) Starting from a language $\mathcal{L}$ with atomic propositions P and Q, build a probability table on its state-descriptions and use that table to prove the fact above.

   (b) Show that Equation (3.49) in Section 3.3 entails that either $cr(Q) = 1$ or $cr(Q \supset P) = 1$.

3.5 Further reading

INTRODUCTIONS AND OVERVIEWS

Alan Hájek (2011a). Conditional Probability. In: *Philosophy of Statistics*. Ed. by Prasanta S. Bandyopadhyay and Malcolm R. Forster. Vol. 7. Handbook of the Philosophy of Science. Amsterdam: Elsevier, pp. 99–136

Describes the Ratio Formula and its motivations. Then works through a number of philosophical applications of conditional probability, and a number of objections to the Ratio Formula. Also discusses conditional-probability-first formalizations (as described in note 3 of this chapter).

Todd A. Stephenson (2000). *An Introduction to Bayesian Network Theory and Usage*. Tech. rep. 03. IDIAP

Section 1 provides a nice, concise overview of what a Bayes Net is and how it interacts with conditional probabilities. (Note that the author uses A, B to express the *conjunction* of A and B.) Things get fairly technical after that as he covers algorithms for creating and using Bayes Nets. Sections 6 and 7, though, contain real-life examples of Bayes Nets for speech recognition, Microsoft Windows troubleshooting, and medical diagnosis.

Christopher R. Hitchcock (2021). Probabilistic Causation. In: *The Stanford Encyclopedia of Philosophy*. Ed. by Edward N. Zalta. Spring 2021

While this entry is primarily about *analyses* of the concept of causation using probability theory, along the way Hitchcock includes impressive coverage of the Principle of the Common Cause, Simpson's Paradox, causal modeling with Bayes Nets, and related material.

CLASSIC TEXTS

Hans Reichenbach (1956). The Principle of Common Cause. In: *The Direction of Time*. Berkeley: University of California Press, pp. 157–60

Text in which Reichenbach introduces his account of common causes in terms of screening off. (Note that Reichenbach uses a period to express conjunction, and a comma rather than a vertical bar for conditional probabilities—what we would write as $cr(A \mid B)$ he writes as $P(B, A)$.)

David David Lewis (1976). Probabilities of Conditionals and Conditional Probabilities. *The Philosophical Review* 85, pp. 297–315

Article in which Lewis presents his triviality argument concerning probabilities of conditionals.

EXTENDED DISCUSSION

Bas C. van Fraassen (1982). Rational Belief and the Common Cause Principle. In: *What? Where? When? Why?* Ed. by Robert McLaughlin. Dordrecht: Reidel, pp. 193–209

Frank Arntzenius (1993). The Common Cause Principle. *PSA: Proceedings of the Biennial Meeting of the Philosophy of Science Association* 2, pp. 227–37

Discuss the meaning and significance of Reichenbach's Principle of the Common Cause, then present possible counterexamples (including counterexamples from quantum mechanics).

Alan Hájek (2011b). Triviality Pursuit. *Topoi* 30, pp. 3–15

Explains the plausibility and significance of the claim that probabilities of conditionals are conditional probabilities. Then canvasses a variety of Lewis-style triviality arguments against that claim.

Notes

1. Here's a good way to double-check that 6 & E is equivalent to 6: Remember that equivalence is mutual entailment. Clearly 6 & E entails 6. Going in the other direction, 6 entails 6, but 6 also entails E. So 6 entails 6 & E. When evaluating conditional credences using the Ratio Formula, we'll often find ourselves simplifying a conjunction down to just one or two of its conjuncts. For this to work, the conjunct that remains has to entail each of the conjuncts that was removed.

2. When I refer to an agent's "credence distribution" going forward, I will often be referring to both her unconditional and conditional credences. Strictly speaking this extends our definition of a "distribution", but since conditional credences supervene on unconditional for rational agents, not much damage will be done.

3. Some authors take advantage of Equation (3.7) to formalize probability theory in exactly the opposite order from the way I've been proceeding. They begin by introducing conditional credences and subject them to a number of constraints somewhat like

Kolmogorov's axioms. The desired rules for *un*conditional credences are then obtained by introducing the single constraint that for all P in $\mathcal{L}$, $cr(P) = cr(P \mid T)$. For more on this approach, its advocates, and its motivations, see Section 5.4.

4. Bayes never published the theorem; Richard Price found it in Bayes's notes and published it after Bayes's death in 1761. Pierre-Simon Laplace independently rediscovered the theorem later on and was responsible for much of its early popularization.

5. In everyday English "likely" is a synonym for "probable". Yet R.A. Fisher introduced the technical term "likelihood" to represent a particular *kind* of probability—the probability of some evidence given a hypothesis. This somewhat peculiar terminology has stuck.

6. Quoted in Galavotti (2005, p. 51).

7. For instance, we have to assume that the base rates of the alleles are equal in the population, none of the relevant phenotypes is fitter than any of the others, and the blue-finned fish don't assortatively mate with other blue-finned fish. (Thanks to Hayley Clatterbuck for discussion.)

8. Throughout this section and Section 3.2.1, I will assume that any proposition appearing in the condition of a conditional cr-expression has a nonzero cr-value. Defining probabilistic independence for propositions with probability 0 can get complicated. (See e.g. Fitelson and Hájek (2014))

9. One will sometimes see "screening off" defined without Equation (3.28) or its analogue. (That is, some authors define screening off in terms of R's making the correlation between P and Q disappear, without worrying whether $\sim R$ has the same effect.) Equation (3.28) makes an important difference to our definition: in the Baxter example I does not screen off H from C according to our definition because when $\sim I$ is supposed, C becomes very relevant to H.

  I have included Equation (3.28) in our definition because it connects our approach to the more general notion of screening off used in the statistics community. In statistics one often works with continuous random variables, and the idea is that random variable Z screens off X from Y if X and Y become independent conditional on each possible value of Z. Understanding proposition R as a dichotomous random variable (Chapter 2, note 7) converts this general definition of screening off into the particular definition I've given in the text.

  Many authors also leave Equation (3.26) (or its analogue) implicit in their definitions of "screening off". But since examples of screening off always involve unconditional correlations that disappear under conditioning, I've made this feature explicit in my definition.

10. Not to be confused with the Rambler's Fallacy: I've said so many false things in a row, the next one must be true!

11. Twenty flips of a fair coin provide a good example of what statisticians call **IID trials**. "IID" stands for "independent, identically distributed." The flips are "independent" because each is probabilistically independent of all the others; information about the outcomes of other coin flips doesn't change the probability that a particular flip will come up heads. The flips are "identically distributed" because each flip has the same probability of producing heads (as contrasted with a case in which some of the flips are of a fair coin while others are flips of a biased coin).

12. This paradoxical phenomenon is named after E.H. Simpson because of a number of striking examples he gave in his (1951). Yet the phenomenon had been known to statisticians as early as Pearson, Lee, and Bramley-Moore (1899) and Yule (1903).

13. I learned about the Harden/DeRozan example from Reuben Stern, who in turn learned about it from a reddit post by a user named Jerome Williams. (I copied the specific data for the two players from stats.nba.com.) The UC Berkeley example was brought to the attention of philosophers by Cartwright (1979).

14. Notice that these findings only address one potential form of bias that might have been present in Berkeley's graduate admissions. For instance, they're consistent with the possibility that women were being actively discouraged from applying to the less selective departments.

15. I'm playing a bit fast and loose with the objects of discussion here. Throughout this chapter we're considering correlations in an agent's credence distribution. Reichenbach was concerned not with probabilistic correlations in an agent's credences but instead with correlations in objective frequencies or chance distributions (about which more in Chapter 5). But presumably if the Principle of the Common Cause holds for objective probability distributions, that provides an agent who views particular propositions as empirically correlated with some reason to suppose that the events described in those propositions either stand as cause to effect or share a common cause.

16. You might worry that Figure 3.3 presents a counterexample to Reichenbach's Principle of the Common Cause, because G and S are unconditionally correlated yet G doesn't cause S and they have no common cause. It's important to the principle that the causal relations need not be *direct*; for Reichenbach's purposes G counts as a cause of S even though it's not the immediate cause of S.

17. Just to indicate a few more complexities that may arise: While our discussion in the text concerns "direct" common causes, one can have an "indirect" common cause that doesn't screen off its effects from each other. For example, if we imagine merging Figures 3.2 and 3.3 to show how the subject's parents' genes are a common cause of both smoking and drinking by way of her addictive personality, it is possible to arrange the numbers so that her parents' genetics don't screen off smoker from drinker. Even more complications arise if some causal arrows do end-arounds past others—what if in addition to the causal structure just described, the parents' genetics tend to make *them* smokers, which in turn directly influences the subject's smoking behavior?

18. Here I assume that a rational agent will entertain an indicative conditional only if she takes its antecedent to be possible. For arguments in favor of this position, and citations to the relevant literature, see Moss (2018, Sect. 4.3) and Titelbaum (2013a, Sect. 5.3.2). The analogous assumption for conditional credences is that an agent assigns a conditional credence only when its condition is true in at least one of her doxastically possible worlds.

19. One *could* study a kind of attitude different from the conditional credences considered in this book—something like a subjunctive degree of belief. Joyce (1999) does exactly that, but is careful to distinguish his analysandum from standard conditional degrees of belief. (For instance, the arguments for the Ratio Formula given earlier in this chapter do not extend to Joycean subjunctive credences.) Schwarz (2018) then evaluates triviality arguments for subjunctive conditional credences much like the triviality arguments for indicatives I will go on to consider in this section.

20. I realize some of the "p"s and "q"s in this quote are flipped around from what one might expect, but that's how it's printed in my copy of Ramsey (1931). Context makes clear that the ordering is *not* why Ramsey rejects the proposed equivalents to "the degree of belief in p given q"; he'd still reject them were the order inverted.

21. A variety of recent positions in linguistics and the philosophy of language suggest that indicative conditionals with modal expressions in their consequents do not obey classical logical rules. Yalcin (2012), among others, classes probability locutions with these modals and so argues that, inter alia, indicative conditionals with probabilistic consequents do not keep *modus tollens* truth-preserving. (His argument could easily be extended to disjunctive syllogism as well.) Yet the alternative positive theory of indicative conditionals Yalcin offers does not analyze conditional credences in terms of conditionals either, so even if he's correct, we would still need an independent understanding of what conditional credences are. (Thanks to Fabrizio Cariani for discussion of these points.)

22. A careful reader will note that the proof given fails when $cr(P \mid Q)$ takes an extreme value. If $cr(P \mid Q) = 0$, the condition in $cr(P \mid Q \& P)$ will have unconditional credence 0, while if $cr(P \mid Q) = 1$, the condition in $cr(P \mid Q \& \sim P)$ will have credence 0. So strictly speaking the triviality result is that (3.48) can hold only when the agent takes P and Q to be independent, or dependent in the strongest possible fashion. This is no more plausible than the less careful version I've given in the text. (Thanks to Glauber de Bona for being the careful reader who caught this for me!)

23. Fitelson (2015) proves a triviality result like Lewis's using probability tables (instead of proceeding axiomatically). Moreover, he traces the triviality specifically to the combination of (3.48) with the assumption that the conditional $\rightarrow$ satisfies what's known as the "import-export" condition.

24. Interestingly, this idea is often traced back to a suggestion in Ramsey, known as "Ramsey's test" (Ramsey 1929/1990, p. 155n).

25. Thanks to Irving Lubliner for inspiring this problem.

26. This problem was inspired by a problem of Branden Fitelson's. Thanks to Catrin Campbell-Moore for devising this particularly elegant set of conditions.

4

Updating by Conditionalization

Up to this point we have discussed *synchronic* credence constraints—rationally required relations among the degrees of belief an agent assigns at a given time. This chapter introduces the fifth (and final) core normative Bayesian rule, Conditionalization. Conditionalization is a *diachronic* rule, requiring an agent's degrees of belief to line up in particular ways across times.

I begin by laying out the rule and some of its immediate consequences. We will then practice applying Conditionalization using Bayes's Theorem. Some of Conditionalization's consequences will prompt us to ask what notions of learning and evidence pair most naturally with the rule. I will also explain why it's important to attend to an agent's *total* evidence in evaluating her responses to learning.

Finally, we will see how Conditionalization helps Bayesians distinguish two influences on an agent's opinions: the content of her evidence, and her tendencies to respond to evidence in particular ways. This will lead to Chapter 5's discussion of whether multiple distinct responses to the same evidence might ever be rationally permissible. Differing answers to that question provide a crucial distinction between Subjective and Objective Bayesianism.

4.1 Conditionalization

Suppose I tell you I just rolled a fair six-sided die, and give you no further information about how the roll came out. Presumably you assign equal unconditional credence to each of the six possible outcomes, so your credence that the die came up six will be 1/6. I then ask you to suppose that the roll came up even (while being very clear that this is just a supposition—I'm still not revealing anything about the actual outcome). Applying the Ratio Formula to your unconditional distribution, we find that rationality requires your credence in six conditional on the supposition of even to be 1/3. Finally, I break down and tell you that the roll actually did come up even. Now how confident should you be that it came up six?

Fundamentals of Bayesian Epistemology 1: Introducing Credences. Michael G. Titelbaum, Oxford University Press.
© Michael G. Titelbaum 2022. DOI: 10.1093/oso/9780198707608.003.0004

I hope the obvious answer is 1/3. When you learn that the die actually came up even, the effect on your confidence in a six is identical to the effect of merely supposing evenness. This relationship between learning and supposing is captured in Bayesians' credence-updating rule:

Conditionalization: For any time t_i and later time t_j, if proposition E in $\mathcal{L}$ represents everything the agent learns between t_i and t_j, and $cr_i(E) > 0$, then for any H in $\mathcal{L}$,

$$cr_j(H) = cr_i(H \mid E)$$

where cr_i and cr_j are the agent's credence distributions at the two times. Conditionalization captures the idea that an agent's credence in H at t_j— after *learning* E—should equal her earlier t_i credence in H had she merely *supposed* E. If we label the two times in the die-roll case t_1 and t_2, and let 6 represent the die's coming up six and E represent its coming up even, then Conditionalization tells us

$$cr_2(6) = cr_1(6 \mid E) \tag{4.1}$$

which equals 1/3 (given what we know about your unconditional distribution at t_1).

Warning

Some theorists take Conditionalization to *define* conditional credence. For them, to assign the conditional credence $cr_i(H \mid E) = r$ *just is* to be disposed to assign $cr_j(H) = r$ should you learn E. As I said in Chapter 3, I take conditional credence to be a genuine mental state, manifested by the agent in various ways at t_i (what she'll say in conversation, what sorts of bets she'll accept, etc.) beyond just her dispositions to update. For us, Conditionalization represents a *normative* constraint relating the agent's unconditional credences at a later time to her conditional credences earlier on.

Combining Conditionalization with the Ratio Formula gives us

$$cr_j(H) = cr_i(H \mid E) = \frac{cr_i(H \& E)}{cr_i(E)} \tag{4.2}$$

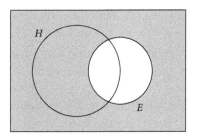

Figure 4.1 Updating on E

(when $cr_i(E) > 0$). A Venn diagram shows why dividing these particular t_i credences should yield the agent's credence in H at t_j. In Chapter 3 we used a diagram like Figure 4.1 to understand conditional credences. There the white circle represented a set of possibilities to which the agent had temporarily narrowed her focus in order to entertain a supposition.

Now let's imagine the rectangle represents all the possible worlds the agent entertains at t_i (her doxastically possible worlds at that time). The size of the H-circle represents the agent's unconditional t_i credence in H. Between t_i and t_j the agent learns that E is true. Among the worlds she had entertained before, the agent now excludes all the non-E worlds. Her set of doxastic possibilities narrows down to the E-circle; in effect, *the E-circle becomes the agent's new rectangle*. How unconditionally confident is the agent in H now? That depends what fraction of her new doxastic space is occupied by H-worlds. And this is what Equation (4.2) calculates: it tells us what fraction of the E-circle is occupied by H & E worlds.

As stated, the Conditionalization rule is useful for calculating a single unconditional credence value after an agent has gained evidence. But what if you want to generate the agent's entire t_j credence distribution at once? We saw in Chapter 2 that a rational agent's entire t_i credence distribution can be specified by a probability table that gives the agent's unconditional t_i credence in each state-description of $\mathcal{L}$. To satisfy the probability axioms, the credence values in a probability table must be non-negative and sum to 1. The agent's unconditional credence in any (non-contradictory) proposition can then be determined by summing her credences in the state-descriptions on which that proposition is true.

When an agent updates her credence distribution by applying Conditionalization to some learned proposition E, we will say that she "conditionalizes on E". (Some authors say she "conditions on E".) To calculate the probability table resulting from such an update, we apply a two-step process:

1. Assign credence 0 to all state-descriptions inconsistent with the evidence learned.
2. Multiply each remaining nonzero credence by the *same* constant so that they all sum to 1.

As an example, let's consider what happens to your confidence that the fair die roll came up prime[1] when you learn that it came up even:

P	E	cr_1	cr_2
T	T	1/6	1/3
T	F	1/3	0
F	T	1/3	2/3
F	F	1/6	0

Here we've used a language $\mathcal{L}$ with atomic propositions P and E representing "prime" and "even". The cr_1 column represents your unconditional credences at time t_1, while the cr_2 column represents your t_2 credences. Between t_1 and t_2 you learn that the die came up even. That's inconsistent with the second and fourth state-descriptions, so in the first step of our update process their cr_2-values go to 0. The cr_1-values of the first and third state-descriptions (1/6 and 1/3 respectively) add up to only 1/2. So we multiply both of these values by 2 to obtain unconditional t_2-credences summing to 1.[2]

In this manner, we generate your unconditional state-description credences at t_2 from your state-description credences at t_1. We can then calculate cr_2-values for other propositions. For instance, adding up the cr_2-values on the lines that make P true, we find that

$$cr_2(P) = 1/3 \qquad (4.3)$$

Given your initial distribution, your credence that the die came up prime after learning that it came up even is required to be 1/3. Hopefully that squares with your intuitions about what's rationally required in this case!

One final note: Our two-step process for updating probability tables yields a handy fact. Notice that in the second step of the process, every state-description that hasn't been set to zero is multiplied by the *same* constant. When two values are multipled by the same constant, the ratio between them remains intact. This means that if two state-descriptions have nonzero credence values after an update by Conditionalization, those values will stand in the same ratio as they did before the update. This fact will prove useful for problem-solving later on. (Notice that it applies only to *state-descriptions*; propositions that are not state-descriptions may not maintain their credence ratios after a conditionalization.)

4.1.1 Consequences of Conditionalization

If we adopt Conditionalization as our updating norm, what follows? When an agent updates by conditionalizing on E, her new credence distribution is just her earlier distribution conditional on E. In Section 3.1.2 we saw that if an agent's credence distribution obeys the probability axioms and Ratio Formula, then the distribution she assigns conditional on any particular proposition (in which she has nonzero credence) will be probabilistic as well. This yields the important result that if an agent starts off obeying the probability axioms and Ratio Formula and then updates by Conditionalization, her resulting credence distribution will satisfy the probability axioms as well.[3]

The process may then iterate. Having conditionalized her probabilistic distribution cr_1 on some evidence E to obtain probabilistic credence distribution cr_2, the agent may then gain further evidence E', which she conditionalizes upon to obtain cr_3 (and so on). Moreover, conditionalization has the elegant mathematical property of being **cumulative**: Instead of obtaining cr_3 from cr_1 in two steps—first conditionalizing cr_1 on E to obtain cr_2, then conditionalizing cr_2 on E' to obtain cr_3—we can generate the same cr_3 distribution by conditionalizing cr_1 on $E \& E'$, a conjunction representing all the propositions learned between t_1 and t_3. (You'll prove this in Exercise 4.3.) Because conditionalization is cumulative it is also **commutative**: Conditionalizing first on E and then E' has the same effect as conditionalizing in the opposite order.

Besides being mathematically elegant, cumulativity and commutativity are intuitively plausible features of a learning process. Suppose a detective investigating a crime learns that the perpetrator was an Italian accordionist, and updates her credences accordingly. Intuitively, it shouldn't matter if we describe this episode as the detective's learning first one piece of evidence and then another (first that the perpetrator was Italian, and then that he was an accordionist) or as the detective's learning a single conjunction containing both. Because conditionalization is cumulative, it will prescribe the same ultimate credences for the detective on either construal. Similarly, it shouldn't matter whether we take her to have learned that the perpetrator was an Italian accordionist or an accordion-playing Italian. Because conditionalization is commutative, the order in which pieces of evidence are presented makes no difference to an agent's ultimate credences.[4]

When an agent conditionalizes on evidence E, what happens to her uncon-ditional credence in that very evidence? Substituting E for H in Equation (4.2) (and recalling that $E \& E$ is equivalent to E), we can see that if an agent learns E between t_i and t_j then

$$cr_j(E) = 1 \qquad (4.4)$$

Conditionalization creates certainties; conditionalizing on a piece of evidence makes an agent certain of that evidence. Moreover, any proposition entailed by that evidence must receive at least as high a credence as the evidence (by our Entailment rule). So an agent who conditionalizes also becomes certain of any proposition entailed by the evidence she learns.

And conditionalization doesn't just create certainties; it also retains them. If an agent is certain of a proposition at t_i and updates by Conditionalization, she will remain certain of that proposition at t_j. That is, if $cr_i(H) = 1$ then Conditionalization yields $cr_j(H) = 1$ as well. On a probability table, this means that once a state-description receives credence 0 at a particular time (the agent has ruled out that possible state of the world), it will receive credence 0 at all subsequent times as well.

In Exercise 4.2 you'll prove that conditionalizing retains certainties from the probability axioms and Ratio Formula. But it's easy to see why this occurs on a Venn diagram. You're certain of H at t_i when H is true in every world you consider a live doxastic possibility. Conditionalizing on E strictly narrows the set of possible worlds you entertain. So if H was true in every world you entertained before conditionalizing, it'll be true in every world you entertain afterwards as well.

Combining these consequences of Conditionalization yields a somewhat counterintuitive result, to which we'll return in later discussions. Condition-alizing on E between two times makes that proposition (and any proposition it entails) certain. Future updates by Conditionalization will then retain that certainty. So if an agent updates by conditionalizing throughout her life, any piece of evidence she learns at any point will remain certain for her ever after.

What if an agent doesn't learn *anything* between two times? Bayesians represent an empty evidence set as a tautology. So when an agent gains no information between t_i and t_j, Conditionalization yields

$$cr_j(H) = cr_i(H \mid T) = cr_i(H) \qquad (4.5)$$

for any H in $\mathcal{L}$. (The latter half of this equation comes from Equation (3.7), which noted that credences conditional on a tautology equal unconditional credences.) If an agent learns nothing between two times and updates by Conditionalization, her degrees of confidence remain unchanged.

4.1.2 Probabilities are weird! The Base Rate Fallacy

Bayes's Theorem expresses a purely synchronic relation; as we saw in Section 3.1.3, for any time t_i it calculates $cr_i(H \mid E)$ in terms of other credences assigned at that time. Nevertheless, our diachronic Conditionalization rule gives Bayes's Theorem added significance. Conditionalization says that your unconditional t_j credence in hypothesis H after learning E should equal $cr_i(H \mid E)$. Bayes's Theorem is a tool for calculating this crucial value (your "posterior" at t_j) from other credences you assign at t_i. As new evidence comes in over time and we repeatedly update by conditionalizing, Bayes's Theorem can be a handy tool for generating new credences from old.

For example, we could've used Bayes's Theorem to answer our earlier question of what happens to your credence in six when you learn that a fair die roll has come up even. The hypothesis is 6, and the evidence is E (for even). By Conditionalization and then Bayes's Theorem,

$$cr_2(6) = cr_1(6 \mid E) = \frac{cr_1(E \mid 6) \cdot cr_1(6)}{cr_1(E)} \tag{4.6}$$

$cr_1(6)$, your prior credence in a six, is $1/6$, and $cr_1(E)$, your prior credence in even, is $1/2$. The likelihood of E, $cr_1(E \mid 6)$, is easy—it's 1. So the numerator is $1/6$, the denominator is $1/2$, and the posterior $cr_1(6 \mid E) = cr_2(6) = 1/3$ as we saw before.[5]

Let's apply Bayes's Theorem to a more interesting case:

> One in 1,000 people have a particular disease. You have a test for the presence of the disease that is 90% accurate, in the following sense: If you apply the test to a subject who has the disease it will yield a positive result 90% of the time, and if you apply the test to a subject who lacks the disease it will yield a negative result 90% of the time.

> You randomly select a person and apply the test. The test yields a positive result. How confident should you be that this subject actually has the disease?

Most people—including trained medical professionals!—answer this question with a value around 80% or 90%. But if you set your credences by the statistics given in the problem, the rationally required degree of confidence that the subject has the disease is less than 1%.

We'll use Bayes's Theorem to work that out. Let D be the proposition that the subject has the disease and P the proposition that when applied to the subject, the test yields a positive result. Here D is our hypothesis, and P is the evidence acquired between t_1 and t_2. At t_1 (before applying the test) we take the subject to be representative of the population, giving us priors for the hypothesis and the catchall:

$$cr_1(D) = 0.001 \qquad cr_1(\sim D) = 0.999$$

The accuracy profile of the test gives us likelihoods for the hypothesis and catchall:

$$cr_1(P \mid D) = 0.9 \qquad cr_1(P \mid \sim D) = 0.1$$

In words, you're 90% confident that the test will yield a positive result given that the subject has the disease, and 10% confident that we'll get a "false positive" on the supposition that the subject lacks the disease.

Now we'll apply a version of Bayes's Theorem from Section 3.1.3, in which the Law of Total Probability has been used to expand the denominator:

$$
\begin{aligned}
cr_2(D) &= \frac{cr_1(P \mid D) \cdot cr_1(D)}{cr_1(P \mid D) \cdot cr_1(D) + cr_1(P \mid \sim D) \cdot cr_1(\sim D)} \\
&= \frac{0.9 \cdot 0.001}{0.9 \cdot 0.001 + 0.1 \cdot 0.999} \\
&\approx 0.009 = 0.9\%
\end{aligned}
\tag{4.7}
$$

So there's the calculation. After learning of the positive test result, your credence that the subject has the disease should be a little bit less than 1%. But even having seen this calculation, most people find it hard to believe. Why shouldn't we be more confident that the subject has the disease? Wasn't the test 90% accurate?

Tversky and Kahneman (1974) suggested that in cases like this one, people's intuitive responses ignore the "base rate" of a phenomenon. The base rate in our example is the prior credence that the subject has the disease. In this case, that base rate is extremely low (1 in 1,000). But people tend to forget about the base rate and be overwhelmed by accuracy statistics (such as likelihoods) about the test. This is known as the **Base Rate Fallacy**.

Why is the base rate so important? To illustrate, let's suppose you applied this test to 10,000 people. Using the base rate statistics, we would expect about ten of those people to have the disease. Since the test gives a positive result for 90% of people who have the disease, we would expect these ten diseased people to yield about nine positive results—so-called "true positives". Then there would be about 9,990 people lacking the disease. Since $cr_i(P \mid \sim D)$—the false positive rate—is 10%, we'd expect to get about 999 false positive results. Out of 1,008 positive results the test would yield, only nine of those subjects (or about 0.9%) would actually have the disease. This particular disease is so rare—its base rate is so tiny—that even with an accurate test we should expect the false positives to swamp the true positives. So when a single randomly selected individual takes the test and gets a positive result, we should be much more confident that this is a false positive than a true one.

Another way to see what's going on is to consider the **Bayes factor** of the evidence you receive in this case. Using Conditionalization and the Ratio Formula, we can derive

$$\frac{cr_j(H)}{cr_j(\sim H)} = \frac{cr_i(H \mid E)}{cr_i(\sim H \mid E)} = \frac{cr_i(H)}{cr_i(\sim H)} \cdot \frac{cr_i(E \mid H)}{cr_i(E \mid \sim H)} \qquad (4.8)$$

That last fraction on the right—the ratio of the likelihood of the hypothesis to the likelihood of the catchall—is the Bayes factor. Personally, I found this equation fairly impenetrable until I remembered that $cr(H)/cr(\sim H)$ is an agent's odds for the proposition H (Section 2.3.4). That means we can rewrite Equation (4.8) as

$$\text{odds for } H \text{ after update} = \text{odds for } H \text{ before update} \cdot \text{Bayes factor} \qquad (4.9)$$

If you update by Conditionalization, learning E multiplies your odds for H by the Bayes factor. The Bayes factor thus provides a handy way to measure how much learning E affects your opinion about the hypothesis.

In our disease example, the Bayes factor is

$$\frac{cr_1(P \mid D)}{cr_1(P \mid \sim D)} = \frac{0.9}{0.1} = 9 \qquad (4.10)$$

At t_1, your odds for D are $1:999$. Applying the test has a substantial influence on these odds; as the Bayes factor reveals, a positive test result multiplies the odds by 9. This reflects the high accuracy of the test. Yet since the odds were so small initially, multiplying them by 9 only brings them up to $9:999$. So even after seeing the test outcome, you should be much more confident that the subject doesn't have the disease than you are that she does.[6]

4.2 Evidence and certainty

Combining Conditionalization with the probability axioms and Ratio Formula creates a Bayesian approach to evidence that many have found troubling. Conditionalization works with a proposition E representing everything the agent learns between two times. (If many propositions are learned, E is their conjunction.) We also speak of E as the evidence the agent gains between those two times. Yet Conditionalization gives E properties that epistemologists don't typically attribute to evidence.

We've already seen that a piece of evidence E (along with anything it entails) becomes certain once conditionalized upon. When an agent learns E, the set of doxastically possible worlds she entertains shrinks to a set of worlds that all make E true; on the Venn diagram, what once was merely an E-circle *within* her rectangle of worlds now becomes the entire rectangle. And as we saw in Section 4.1.1, this change is permanent: as long as the agent keeps updating by Conditionalization, any evidence she once learned remains certain and possible worlds inconsistent with it remain ruled out.

Is there any realistic conception of evidence—and of learning—that satisfies these conditions? When I learn that my sister is coming over for Thanksgiving dinner, I become highly confident in that proposition. But do I become 100% certain? Do I *rule out* all possible worlds in which she doesn't show, refusing to consider them ever after? As Richard C. Jeffrey put it:

> Certainty is quite demanding. It rules out not only the far-fetched uncertainties associated with philosophical skepticism, but also the familiar uncertainties that affect real empirical inquiry in science and everyday life.
>
> (2004, p. 53)

This concern about certainties motivates the

Regularity Principle: In a rational credence distribution, no logically contingent proposition receives unconditional credence 0.

The Regularity Principle captures the common-sense idea that one's evidence is never so strong as to entirely rule out any logical possibility. (Recall that a logically contingent proposition is neither a logical contradiction nor a logical tautology.)[7] As damning evidence against a contingent proposition mounts up, we may keep decreasing and decreasing our credence in it, but our unconditional credence distribution should always remain **regular**—it should assign each logically contingent proposition at least a tiny bit of confidence.[8]

The Regularity Principle adds to the synchronic Bayesian rules we have seen so far—it is not entailed by the probability axioms, the Ratio Formula, or any combination of them. As our Contradiction result showed in Section 2.2.1, those rules do entail that all logical contradictions receive credence 0. But Regularity is the converse of Contradiction; instead of saying that *all* contradictions receive credence 0, it entails that *only* contradictions do. Similarly, Regularity (along with the probability axioms) entails the converse of Normality: instead of saying that *all* tautologies receive credence 1, it entails that *only* tautologies do. (The negation of a contingent proposition is contingent; if we were to assign a contingent proposition credence 1 its negation would receive credence 0, in violation of Regularity.) This captures the common-sense idea that one should never be absolutely certain of a proposition that's not logically true.[9]

Conditionalization conflicts with Regularity; the moment an agent conditionalizes on contingent evidence, she assigns credence 1 to a non-tautology. As we saw earlier, conditionalizing on contingent evidence rules out doxastic possibilities the agent had previously entertained; on the Venn diagram, it narrows the set of worlds under consideration. Regularity, on the other hand, fixes an agent's doxastic possibility set as the full set of logical possibilities. While evidence might shift the agent's credences around among various possible worlds, an agent who satisfies Regularity will never eliminate a possible world outright.

We might defend Conditionalization by claiming that whenever agents receive contingent evidence, it is of a highly specific kind, and Regularity is false for this kind of evidence. Perhaps I don't actually learn that my sister is coming over for Thanksgiving—I learn that she *told* me she's coming; or that it *seemed* to me that she said that; or that I had a phenomenal experience as of…. Surely I can be certain what phenomenal experiences I've had, or at least what experiences I'm having right now. While in the midst of having a particular phenomenal experience, can't I entirely rule out the logical possibility that I am having a different experience instead? C.I. Lewis defended this approach as follows:

> If anything is to be probable, then something must be certain. The data which themselves support a genuine probability, must themselves be certainties. We do have such absolute certainties, in the sense data initiating belief and in those passages of experience which later may confirm it.   (1946, p. 186)

Yet foundationalist epistemologies based on sense data and indubitable phenomenology have become unpopular in recent years. So it's worth considering other ways to make sense of Conditionalization's conception of evidence. Levi (1980) took credence-1 propositions to represent "standards of serious possibility":

> When witnessing the toss of a coin, [an agent] will normally envisage as possibly true the hypothesis that the coin will land heads up and that it will land tails up. He may also envisage other possibilities—e.g., its landing on its edge. However, if he takes for granted even the crudest folklore of modern physics, he will rule out as impossible the coin's moving upward to outer space in the direction of Alpha Centauri. He will also rule out the hypothesis that the Earth will explode.   (p. 3)

However, Levi formalized his standards of serious possibility so that they could change—growing either stronger or weaker—for a given agent over time. So his approach did not fully embrace Conditionalization.

Alternatively, we could represent agents as ruling out contingent possibilities only relative to a particular inquiry. Consider a scientist who has just received a batch of experimental data and wants to weigh its import for a set of hypotheses. There are always outlandish possibilities to consider: the data might have been faked; the laws of physics might have changed a moment ago; she might be a brain in a vat. But to focus on the problem at hand, she might conditionalize on the data and see where that takes her credences in the hypotheses. Updating by Conditionalization might fail as a big-picture, permanent strategy, but nevertheless could be useful in carefully delimited contexts. (I mentioned this possibility in Section 2.2.3.)

Perhaps these interpretations of evidence conditionalized-upon remain unsatisfying. We will return to this problem in Chapter 5, and consider a generalized updating rule (Jeffrey Conditionalization) that allows agents to redistribute their credences over contingent possibilities without eliminating any of them entirely. For the rest of this chapter we will simply assume that Conditionalization on some kind of contingent evidence is a rational updating rule, so as to draw out further features of such updates.

4.2.1 Probabilities are weird! Total Evidence and the Monty Hall Problem

Classical entailment is **monotonic** in the following sense: If a piece of evidence *E* entails *H*, any augmentation of that evidence (any conjunction that includes *E* as a conjunct) will entail *H* as well. Probabilistic relations, however, can be

nonmonotonic: H might be highly probable given E, but improbable given $E \& E'$. For this reason, it's important for an agent assigning credences on the basis of her evidence to consider *all* of that evidence, and not simply draw conclusions from a subset of it. Carnap (1947) articulated the **Principle of Total Evidence** that a rational agent's credence distribution takes into account all of the evidence she possesses.

An agent may violate the Principle of Total Evidence by failing to take into account the *manner* in which she gained particular information. If the agent is aware of the mechanism by which a piece of information was received, it can be important to recognize facts about that mechanism as a component of her total evidence (along with the information itself). In Eddington's (1939) classic example, you draw a sample of fish from a lake, and all the fish you draw are longer than six inches. Normally, updating on this information would increase your confidence that every fish in the lake is at least that long. But if you know the net used to draw the sample has big holes through which shorter fish fall, a confidence increase is unwarranted. Here it's important to conditionalize not only on the lengths of the fish but also on how they were caught. The method by which your sample was collected has introduced an **observation selection effect** into the data.[10]

Observation selection effects are crucial to a famously counterintuitive probability puzzle, the **Monty Hall Problem** (Selvin 1975):

> In one of the games played on *Let's Make a Deal*, a prize is randomly hidden behind one of three doors. The contestant selects one door, then the host (Monty Hall) opens one of the doors the contestant didn't pick. Monty knows where the prize is, and makes sure to always open a door that doesn't have the prize behind it. (If both the unselected doors are empty, he randomly chooses which one to open.) After he opens an empty door, Monty asks the contestant if she wants what's behind the door she initially selected, or what's behind the other remaining closed door. Assuming she understands the details of Monty's procedure, how confident should the contestant be that the door she initially selected contains the prize?

Most people's initial reaction is to answer 1/2: the contestant originally spread her credence equally among the three doors; one of them has been revealed to be empty; so she should be equally confident that the prize is behind each of the remaining two. This analysis is illustrated by the following probability table:

	cr_1	cr_2
Prize behind door A	1/3	1/2
Prize behind door B	1/3	0
Prize behind door C	1/3	1/2

Here we've used the obvious partition of three locations where the prize might be. Without loss of generality, I've imagined that the contestant initially selects door A and Monty then opens door B. At time t_1—after the contestant has selected door A but before Monty has opened anything—she is equally confident that the prize is hidden behind each of the three doors. When Monty opens door B at t_2, the contestant should conditionalize on the prize's not being behind that door. This yields the cr_2 distribution listed above, which matches most people's intuitions.

Yet the contestant's *total* evidence at t_2 includes not only the fact that the prize isn't behind door B but also the fact that Monty chose that door to open for her. These two propositions aren't equivalent among the agent's doxastically possible worlds at t_1; there are possible worlds consistent with what the contestant knows about Monty's procedure in which door B is empty yet Monty opens door C. That door B was not only empty but was *revealed* to be empty is not expressible in the partition used above. So we need a richer partition, containing information both about the location of the prize and about what Monty does:

	cr_1	cr_2
Prize behind door A & Monty reveals B	1/6	1/3
Prize behind door A & Monty reveals C	1/6	0
Prize behind door B & Monty reveals C	1/3	0
Prize behind door C & Monty reveals B	1/3	2/3

Given what the agent knows of Monty's procedure, these four propositions partition her doxastic possibilities at t_1. At that time she doesn't know where the prize is, but she has initially selected door A (and Monty hasn't opened anything yet). If the prize is indeed behind door A, Monty randomly chooses whether to open B or C. So the contestant divides her 1/3 credence that the prize is behind door A equally between those two options. If the prize is behind door B, Monty is forbidden to open that door as well as the door the contestant selected, so Monty must open C. Similarly, if the prize is behind door C, Monty must open B.

At t_2 Monty has opened door B, so the contestant conditionalizes by setting her credences in the second and third partition members to 0, then multiplying the remaining values by a constant so that they sum to 1. This maintains the ratio between her credences on the first and fourth lines; initially she was twice as confident of the fourth as the first, so she remains twice as confident after the update. She is now 2/3 confident that the prize isn't behind the door she initially selected, and 1/3 confident that her initial selection was correct. If she wants the prize, the contestant should switch doors.

This is the correct analysis. If you find that surprising, the following explanation may help: When the contestant originally selected her door, she was 1/3 confident that the prize was behind it and 2/3 confident that the prize was somewhere else. If her initial pick was correct, she claims the prize just in case she sticks with that pick. But if her initial selection was wrong, she wins by switching to the other remaining closed door, because it must contain the prize. So there's a 1/3 chance that sticking is the winning strategy, and a 2/3 chance that switching will earn her the prize. Clearly switching is a better idea!

When I first heard the Monty Hall Problem, even that explanation didn't convince me. I only became convinced after I simulated the scenario over and over and found that sticking made me miss the prize roughly two out of three times. If you're not convinced, try writing a quick computer program or finding a friend with a free afternoon to act as Monty Hall for you a few hundred times. You'll eventually find that the probability table taking *total* evidence into account provides the correct analysis.[11]

One final note about total evidence: I may have convinced you that taking your total evidence into account is a good idea, but you might be concerned that it's impossible. After all, at each conscious moment an agent receives torrents of information from her environment. How can she take it *all* into account when assigning a credence to a particular proposition—say, the proposition that the cheese sandwich on the counter in front of her has not yet gone bad?

The nonmonotonicity of probabilistic relations means that a rational agent cannot afford to ignore any of her evidence. But many of the propositions an agent learns in a given moment will be *irrelevant* to the matter under consideration relative to her current credence distribution. That is, for many pieces of evidence her credence in the proposition at issue would be the same whether she conditionalized on that particular piece of evidence or not. As the agent ponders her cheese sandwich, information about the color of the bird that just flew by or the current position of her right hand makes no difference to her credence that the sandwich is edible. So while a rational agent doesn't *ignore* any of her total evidence, the irrelevance of much of that evidence permits her to focus in on the few pieces of evidence that are relevant to the proposition under consideration. For this reason, Bayesians often bypass discussion of an agent's total evidence in favor of discussing her total *relevant* evidence.[12]

4.3 Priors and standards

4.3.1 Initial priors

Consider a rational agent with probabilistic credences who updates by Conditionalization each time she gains new evidence, for her entire life. At a given moment t_i she has a credence distribution cr_i. She then gains new evidence E and updates by Conditionalization. Her unconditional cr_i values provide the priors for that update, and her cr_i values conditional on E provide the posteriors. By Conditionalization, these posteriors become her unconditional credences at the next time, t_j. Then she receives a new piece of evidence E'. Her unconditional cr_j values supply the priors for a new update, and her cr_j values conditional on E' are the posteriors.

And so it goes. We have already seen that if this agent updates by Conditionalization *every* time she learns something new, she will gain contingent certainties over time and never lose any of them. So her entire doxastic life will be a process of accumulating empirical evidence from her environment, building a snowball of information that never loses any of its parts.

What happens if we view that process backwards, working from the agent's present doxastic state back through the states she assigned in the past? Her current unconditional credences resulted from an earlier update by Conditionalization. Relative to that update, her current credences were the posteriors and some other distribution provided the priors. But those priors, in turn, came from a conditionalization. So they were once the posteriors of an even *earlier* set of priors. As we go backwards in time, we find a sequence of credence distributions, each of which was conditionalized to form the next. And since each conditionalization strictly added evidence, the earlier distributions contain successively less and less contingent information as we travel back.

Bayesian epistemologists often imagine marching backwards in this fashion until there's no farther back to go. They imagine that if you went back far enough, you would find a point at which the agent possessed literally no contingent information. This was the starting point from which she gained her

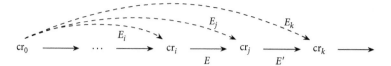

Figure 4.2 An initial prior?

very first piece of evidence, and made her first update by Conditionalization. The agent's credence distribution at this earliest point is sometimes called her **initial prior distribution** (or her "**ur-prior**").

Let's think about the properties an initial prior distribution would have. First, since the credence distributions that develop from an initial prior by Conditionalization are probability distributions, it's generally assumed that the initial prior satisfies the Kolmogorov axioms (and Ratio Formula) as well. Second, it's thought that since at the imagined initial moment (call it t_0) the agent possessed no contingent information, she should not have been certain of any contingent propositions. In other words, the initial prior distribution cr_0 should be regular (should assign nonextreme values to all contingent propositions). Finally, think about how cr_0 relates to a credence distribution our agent assigns at some arbitrary moment t_i later on. We could recover cr_i by conditionalizing cr_0 on the first piece of evidence the agent ever learned, then conditionalizing the result of that update on the second piece of evidence she learned, and so on until we reach cr_i. But since conditionalizing is cumulative, we could also roll up together all of these intermediate steps and get from cr_0 to cr_i in one move. Suppose the proposition E_i represents the agent's total evidence at t_i—the conjunction of all the individual pieces of evidence she's learned since t_0. Then as long as the agent has updated by conditionalizing at every step between t_0 and t_i, cumulativity guarantees that $cr_i(\cdot) = cr_0(\cdot \mid E_i)$. A rational agent's credence distribution at any given time is her initial prior distribution conditional on her total evidence at that time.

This idea is illustrated in Figure 4.2. Each credence distribution is generated from the previous one by conditionalizing on the evidence learned (solid arrows). But we can also derive each distribution directly (dashed arrows) by conditionalizing cr_0 on the agent's total evidence at the relevant time (E_i for cr_i, E_j for cr_j, etc.).

The initial-priors picture is an attractive one, and bears a certain mathematical elegance. The trouble is that it can at best be a myth. Was there ever a time in a real agent's life when she possessed *no* contingent information? Since cr_0 satisfies the probability axioms, it must be perfectly attuned to logical relations (such as mutual exclusivity and entailment) and assign a value of 1 to all tautologies. So the agent who assigns this initial prior must be omniscient logically while totally ignorant empirically. In seminars, David Lewis used to call such highly intelligent, blank creatures "superbabies"; while some Bayesian artificial intelligence systems may be like this, I doubt any human has ever been.[13] Moreover, I'm not sure it even makes sense for an agent with no contingent information to assign precise numerical credences to the kinds of elaborate, highly detailed empirical claims that are real humans' stock in trade.

4.3.2 Epistemic standards

Yet the formal mechanism employed by the initial priors myth—a regular probability distribution conditionalized on total evidence to generate credence distributions at arbitrary times—can be repurposed to represent something important in epistemology. To get a sense of what I'm talking about, let's consider an example:

Question: When playing a hand of five-card stud, how confident should you become that you'll win the hand upon learning that your last card will be the two of clubs?

Answer: Depends how far you are in the game, and what's happened up to that point.

Five-card stud is a poker game in which you receive a total of five cards, one at a time. Four of a kind (four out of five cards showing the same number) is an excellent, almost unbeatable hand in this game. So let's suppose that your first four cards in this particular hand of five-card stud were the jack of spades, the two of diamonds, the two of hearts, and then the two of spades. With that background information, discovering that your last card will be the two of clubs should make you almost certain that you'll win. (Depending in part on what you know of the other players' hands.)

An agent's **epistemic standards** govern how she reacts to particular pieces of news. These epistemic standards are determined in part by an agent's total evidence, and as such evolve over time. At the beginning of a hand of five-card stud, before any cards are dealt, learning that your last card will be the two of clubs (perhaps by peeking into the deck) would not make you very confident of winning the hand. Similarly, after seeing your first card (the jack of spades), a final two of clubs wouldn't seem like very good news. But for each successive two you receive after that point, your ongoing epistemic standards change such that learning the final card will be a two would make you more and more confident of a win.

When two people react differently to acquiring the same piece of information, they are applying different epistemic standards. We usually attribute the difference in their standards to differences in their previous experience. When one student in a class insists on answering every question, pontificates at length, and refuses to consider others' ideas, some of his fellow students might conclude that this person is the most knowledgeable in the room. But the teacher (or other students with more experience) might draw the opposite conclusion, informed by a broader pool of evidence about how particular personality types behave in conversation.

Yet how should we understand cases in which agents draw different conclusions despite sharing the same *total* evidence? Hiring committees form different beliefs about candidates' suitability from the same application files; jurors disagree about a defendant's guilt after witnessing the same trial; scientists embrace different hypotheses consistent with the same experimental data. These seem to be cases in which agents share a common body of total evidence, or at least total evidence *relevant* to the question at hand. So it can't be some further, unshared piece of extra evidence that's leading the agents to draw differing conclusions.

One could stubbornly maintain that in *every* real-life case in which agents interpret a piece of evidence differently, that difference is *entirely* attributable to the vagaries of their background information. But I think this would be a mistake. In addition to variations in their total evidence, agents have varying ways of interpreting their total evidence. Some people are naturally more skeptical than others, and so require more evidence to become confident of a particular proposition (that humans actually landed on the moon, that a lone gunman shot JFK, that the material world exists). Some people are interested in avoiding high confidence in falsehoods, while others are more interested in achieving high confidence in truths. (The former will tend to prefer noncommittal credences, while the latter will be more willing to adopt credence values near 0 and 1.) Some scientists are more inclined to believe elegant theories, while others incline toward the theory that hews closest to the data. (When the Copernican theory was first proposed, heliocentrism fit the available astronomical data *worse* than Ptolemaic approaches (Kuhn 1957).)

The five-card stud example illustrated what I will call *ongoing* epistemic standards. Ongoing epistemic standards reflect how an agent is disposed at a given time to assign attitudes in light of particular pieces of evidence she might receive. At any given time, an agent's credences can be determined from the last piece of evidence she acquired and the ongoing standards she possessed just before she acquired it (with the latter having been influenced by pieces of evidence she acquired even earlier than *that*). Yet there's another way to think about the influences on an agent's attitudes at a given time: we can separate out the influence of her *total* evidence from the influence of whatever additional, non-evidential factors dictate how she assesses that evidence. I refer to the latter as the agent's *ultimate* epistemic standards. An agent's ultimate epistemic standards capture her evidence-independent tendencies to respond to whatever package of total evidence might come her way.[14]

Warning

Don't take the talk of "standards" in "epistemic standards" too literally. It's not as if agents have a clearly defined set of *principles* that they apply to evidence whenever they interpret its significance, each of which is "an epistemic standard". Most of the time we draw conclusions from evidence without much deliberate thought at all, through a variety of methods that are scarcely systematic. Still, epistemologists need a way to refer to an agent's entire, disorganized bundle of dispositions to interpret evidence in particular ways—that bundle, taken as a unified whole, is what we mean by "her epistemic standards".[15]

In the next section we'll develop a formal, Bayesian representation of an agent's epistemic standards. Again, don't read the math as a report of what's going on in the agent's psychology—I don't do a bunch of probability calculations every time I try to figure out where my earbuds are. Instead, think of it as a succinct way to summarize what the agent would conclude from a variety of bodies of total evidence, regardless of how the agent actually goes about drawing such conclusions.

4.3.3 Hypothetical priors

How might we formally represent an agent's ultimate epistemic standards? A moment ago, I said that the attitudes an agent adopts at any given time combine two influences: her total evidence and her ultimate epistemic standards. So we can think of the agent's epistemic standards as a function from possible bodies of evidence to sets of attitudes adopted in response. This function can be represented in many ways—as a table listing outputs for specific inputs, as a complicated graph, etc. But Bayesians have a particularly useful representation already to hand. We can represent an agent's ultimate epistemic standards using a regular probability distribution Pr_H over her language $\mathcal{L}$, which we call her **hypothetical prior distribution**. I'll presently give an example of how to build such a representation from information about an agent's credences. The key point for now is that once we've constructed a hypothetical prior for an agent, we can recover from it her credence distribution cr_i at any given time t_i, as long as we know her total evidence E_i at that time. We simply conditionalize Pr_H on E_i, and the result is cr_i. Combining the agent's ultimate epistemic standards (Pr_H) with her total evidence (E_i) recovers her set of attitudes (cr_i).

Once we've constructed a hypothetical prior Pr_H for an agent, we can conditionalize it not only on bodies of total evidence she's possessed in the past but also on bodies of evidence she might possess in the future. The hypothetical prior is thus a highly efficient summary of all the attitudes an agent might ever adopt in response to bodies of evidence at various times of her life. Almost miraculously, such a compact summary will be available for any agent whose lifelong credences satisfy the five core Bayesian rules. This is guaranteed by the

Hypothetical Priors Theorem: Given any finite series of credence distributions $\{\text{cr}_1, \text{cr}_2, \ldots, \text{cr}_n\}$, each of which satisfies the probability axioms and Ratio Formula, let E_i be a conjunction of the agent's total evidence at t_i. If each cr_i is related to cr_{i+1} as specified by Conditionalization, then there exists at least one regular probability distribution Pr_H such that for all $1 \leqslant i \leqslant n$,

$$\text{cr}_i(\cdot) = \text{Pr}_H(\cdot \mid E_i)$$

In other words, if at each time in an agent's life her credence distribution satisfies the probability axioms and Ratio Formula, and if she updates those distributions from one time to the next according to Conditionalization, then there will necessarily exist at least one hypothetical prior Pr_H that relates to every credence distribution she ever assigns according to the formula above. (You'll prove this theorem in Exercise 4.8.)

Notice that the Hypothetical Priors Theorem specifies the existence of a *regular* probability distribution Pr_H. Given any rational Bayesian agent, we can recover her credence distribution at a given time by conditionalizing her hypothethical prior on her total contingent evidence at that time. Yet being regular, the hypothetical prior does not assign any contingent certainties itself.

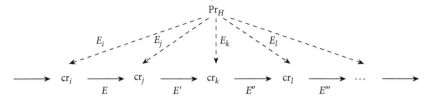

Figure 4.3 A hypothetical prior

So when we are confronted with the agent's credences at a particular time, we can cleanly factor out the two distinct influences on those credences: her total evidence (a set of contingent propositions in her language she takes for certain at that time) and her epistemic standards (represented by a hypothetical prior distribution assigning no contingent certainties). This is how a Bayesian represents the idea that ultimate epistemic standards are extra-evidential; instead of containing any contingent information about the world, epistemic standards express whatever the agent brings to bear on her evidence that isn't based on evidence itself.[16]

Hypothetical priors are convenient because they have a mathematical form with which we're already familiar: a regular probability distribution over language $\mathcal{L}$. Yet a hypothetical prior distribution is not a *credence* distribution.[17] An agent's hypothetical priors are not degrees of belief we imagine she espoused at some particular point in her life, or would espouse under some hypothetical conditions. This is what distinguishes them from the mythical initial priors.[18] Hypothetical priors summarize an agent's abstract evidential assessment tendencies, and stay constant throughout her life as long as she obeys the Conditionalization update rule. Instead of appearing somewhere *within* the series of credence distributions the agent assigns over time, the hypothetical prior "floats above" that series, combining with the agent's total evidence to create members of the series at each given time. As de Finetti puts it, "If we reason according to Bayes' theorem we do not change opinion. We keep the same opinion and we update it to the new situation" (de Finetti 1995, p. 100, translated by and quoted in Galavotti 2005, p. 215).

This arrangement is depicted in Figure 4.3. Again, the solid arrows represent conditionalizations from one time to the next, while the dashed arrows represent the possibility of generating an agent's distribution at any given time by conditionalizing Pr_H on her total evidence at that time.

Now let's show how we can construct a hypothetical prior from information about the credences an agent assigns over time. Suppose Ava has drawn two coins from a bin that contains only fair coins and coins biased toward heads.

Prior to time t_1 she has inspected both coins and determined them to be fair. Between t_1 and t_2 she flips the first coin, which comes up heads. Between t_2 and t_3, the second coin comes up tails.

Our language $\mathcal{L}$ will contain three atomic propositions: N, that neither coin Ava picked is biased; Ha, that the first flip comes up heads; and Hb, that the second flip is heads. Presumably the following probability table describes Ava's credences over time:

N	Ha	Hb	cr_1	cr_2	cr_3
T	T	T	1/4	1/2	0
T	T	F	1/4	1/2	1
T	F	T	1/4	0	0
T	F	F	1/4	0	0
F	T	T	0	0	0
F	T	F	0	0	0
F	F	T	0	0	0
F	F	F	0	0	0

In this example, Ava's total evidence at t_1 (or at least her total evidence representable in language $\mathcal{L}$) is N. We'll call this proposition E_1. Between t_1 and t_2, Ava learns Ha. So cr_2 is cr_1 conditionalized on Ha, and Ava's total evidence at t_2 (which we'll call E_2) is $N \& Ha$. After t_2 Ava learns $\sim Hb$, so cr_3 is cr_2 conditionalized on $\sim Hb$, and E_3 is $N \& Ha \& \sim Hb$. Notice that since N is part of Ava's evidence at all times reflected in this table, she assigns credence 0 throughout the table to any state-description on which N is false.

Since Ava's credence distributions cr_1 through cr_3 are probabilistic, and update by Conditionalization, the Hypothetical Priors Theorem guarantees the existence of at least one hypothetical prior Pr_H standing in a particular relation to Ava's credences. I've added a column to the probability table below representing one such hypothetical prior:

N	Ha	Hb	cr_1	cr_2	cr_3	Pr_H
T	T	T	1/4	1/2	0	1/16
T	T	F	1/4	1/2	1	1/16
T	F	T	1/4	0	0	1/16
T	F	F	1/4	0	0	1/16
F	T	T	0	0	0	21/64
F	T	F	0	0	0	11/64
F	F	T	0	0	0	11/64
F	F	F	0	0	0	5/64

As the Hypothetical Priors Theorem requires, Pr_H is a probability distribution (the values in the Pr_H column are non-negative and sum to 1), and it's regular (no contingent proposition receives a 0). Pr_H stands in the desired relationship to each of cr_1, cr_2, and cr_3: each of those distributions can be obtained from Pr_H by conditionalizing on Ava's total evidence at the relevant time. To take one example, consider cr_2. E_2 is $N \& Ha$. To conditionalize Pr_H on $N \& Ha$, we write a zero on each line whose state-description is inconsistent with $N \& Ha$. That puts zeroes on the third through eighth lines of the table. We then multiply the Pr_H values on the first and second lines by a constant (in this case, 8) so that the results sum to 1. This yields the cr_2 distribution in the table. With a bit of work you can verify that cr_1 results from conditionalizing Pr_H on E_1, and cr_3 is the result of conditionalizing Pr_H on E_3.

The hypothetical prior I wrote down isn't unique. I could have written down (infinitely) many other regular, probabilistic distributions that stand in the required relation to cr_1 through cr_3. This reveals that the information in the original table underdescribes Ava's ultimate epistemic standards, even over our fairly limited language $\mathcal{L}$. For instance, the original table doesn't tell us what credences Ava would've assigned had she learned before t_1 that at least one of the coins was biased. The Pr_H I've provided makes very specific assumptions about Ava's tendencies for that case. (For a fun exercise, figure out what that Pr_H assumes about the biased coins in the bin.) But I could've made different assumptions, and generated a different hypothetical prior consistent with cr_1 through cr_3.

Interestingly, those assumptions don't matter much for practical purposes. Suppose you're working with the series of credence distributions an agent assigned up to some designated time. Typically there will be multiple hypothetical priors consistent with that series of distributions. Those hypothetical priors will differ on what attitudes the agent would have assigned had she received different bodies of total evidence prior to the designated time. But given the course of evidence the agent actually received and series of credence distributions she actually assigned, every hypothetical prior consistent with that series of distributions will make the same predictions about how she will respond to particular pieces of evidence after the designated time. (Assuming she continues to conditionalize.) For instance, every hypothetical prior consistent with Ava's cr_1 and cr_2 distributions will predict the same response to her learning $\sim Hb$ between t_2 and t_3. So which of the available hypothetical priors we use to represent a particular agent's epistemic standards turns out to be irrelevant going forward.

On the other hand, when distinct *agents* have different hypothetical priors, those differences can be important. Plugging an agent's total evidence at a given time into her hypothetical prior yields her credence distribution at that time. When two agents have different hypothetical priors, plugging in the same body of total evidence may yield different results. So two agents may assign different credences to the same proposition in light of the same total evidence. The difference in hypothetical priors is a Bayesian's way of representing that these agents interpret evidence differently, and so may draw different conclusions from the same total body of information.

The obvious next question is whether they can both be *rational* in doing so. Evidence and epistemic standards come up in a variety of contexts in epistemology. As we've just seen, Bayesian epistemology provides an elegant formal apparatus for separating out these two elements. But once we've isolated them, the next question to ask about ultimate epistemic standards is whether anything goes. Is *any* hypothetical prior rational, so long as it's probabilistic? Some probabilistic hypothetical priors will be anti-inductive, or will recommend highly skeptical attitudes in the face of everyday bodies of total evidence. Can we rule out such hypothetical priors as rationally impermissible? Can we go even farther than that, laying down enough rational constraints on ultimate epistemic standards so that any time two agents interpret the same evidence differently, at least one of them must be interpreting it irrationally? This will be our first topic in Chapter 5, as we distinguish Objective from Subjective Bayesianism.

4.4 Exercises

Unless otherwise noted, you should assume when completing these exercises that the credence distributions under discussion satisfy the probability axioms and Ratio Formula. You may also assume that whenever a conditional credence expression occurs or a proposition is conditionalized upon, the needed proposition has nonzero unconditional credence so that conditional credences are well-defined.

Problem 4.1. ☽ Galileo intends to determine whether gravitational acceleration is affected by mass by dropping two cannonballs with differing masses off the Leaning Tower of Pisa. Conditional on there being no effect, he is 95% confident that the cannonballs will land within 0.1 seconds of each other. (The experiment isn't perfect—one ball might hit a bird.) Conditional on mass affecting acceleration, he is 80% confident that the balls *won't* land within 0.1 seconds of each other. (There's some chance that although mass affects acceleration, it doesn't have *much* of an effect.)[19]

(a) Before performing the experiment, Galileo is 30% confident that mass does not affect acceleration. How confident is he that the cannonballs will land within 0.1 seconds of each other?

(b) After Galileo conditionalizes on the evidence that the cannonballs landed within 0.1 seconds of each other, how confident is he in each hypothesis?

Problem 4.2. 🌶 Prove that conditionalizing retains certainties. In other words, prove that if $cr_i(H) = 1$ and cr_j is generated from cr_i by Conditionalization, then $cr_j(H) = 1$ as well.

Problem 4.3. 🌶🌶 Prove that conditionalization is cumulative. That is, prove that for any cr_i, cr_j, and cr_k, conditions 1 and 2 below entail condition 3.

1. For any proposition X in $\mathcal{L}$, $cr_j(X) = cr_i(X | E)$.
2. For any proposition Y in $\mathcal{L}$, $cr_k(Y) = cr_j(Y | E')$.
3. For any proposition Z in $\mathcal{L}$, $cr_k(Z) = cr_i(Z | E \,\&\, E')$.

Problem 4.4. 🌶🌶

(a) Provide a real-life example in which an agent conditionalizes on new evidence, yet her credence in a proposition *compatible with* the evidence decreases. That is, provide an example in which H and E are consistent, yet $cr_2(H) < cr_1(H)$ when E is learned between t_1 and t_2.

(b) Prove that when an agent conditionalizes on new evidence, her credence in a proposition that *entails* the evidence cannot decrease. That is, when $H \vDash E$, it must be the case that $cr_2(H) \geq cr_1(H)$ when E is learned between t_1 and t_2.

(c) Prove that as long as $cr_1(H)$ and $cr_1(E)$ are both nonextreme, conditionalizing on E increases the agent's credence in H when $H \vDash E$.[20]

Problem 4.5. 🌶🌶 Reread the details of the Base Rate Fallacy example in Section 4.1.2. After you apply the diagnostic test once and it yields a positive result, your odds for D should be $9 : 999$.

(a) Suppose you apply the test a second time to the same subject, and it yields a positive result once more. What should your odds for the subject's having the disease be now? (Assume that D screens off the results of the first test from the results of the second.)

(b) How many consecutive tests (each independent of all the prior test results conditional on both D and $\sim D$) would have to yield positive results before your confidence that the subject has the disease exceeded 50%?

(c) Does this shed any light on why patients diagnosed with rare diseases are often advised to seek a second opinion? Explain.

Problem 4.6. Your friend Jones is a gambler. He even gambles about whether to gamble! Every time he goes to the track, he flips a fair coin to determine whether to bet that day. If it comes up heads he bets on his favorite horse, Speedy. If it comes up tails he doesn't bet at all.

On your way to the track today, you were 1/6 confident that out of the six horses running, Speedy would win. You were 1/2 confident that Jones's coin would come up heads. And you considered the outcome of the horse race independent of the outcome of the coin flip. But then you saw Jones leaving the track with a smile on his face. The smile tells you that either Jones bet on Speedy and won, or Jones didn't bet and Speedy didn't win.[21]

(a) 🌙 Using a language with the atomic propositions H (for heads on the coin) and S (for a Speedy win), express the information you learn when you see Jones smiling.

(b) 🌙 After updating on this information by conditionalizing, how confident are you that Speedy won? How confident are you that the coin came up heads?

(c) 🌙🌙🌙 Explain why one of the unconditional credences you calculated in part (b) differs from its prior value and the other one doesn't. Be sure to include an explanation of why *that* unconditional credence was the one that changed out of the two. ("Because that's what the math says" is not an adequate explanation—we want to know why the mathematical outcome *makes sense*.)

(d) 🌙 Evaluate the following claim: "If an agent starts off viewing proposition X as irrelevant to proposition Y, then learning that X and Y have the same truth-value shouldn't change that agent's credence in Y." Explain your answer.

Problem 4.7. 🌙🌙 At t_1, t_2, and t_3, Jane assigns credences over the language $\mathcal{L}$ constructed from atomic propositions P and Q. Jane's distributions satisfy constraints 1 through 6:

1. At t_1, Jane is certain of $Q \supset P$, anything that proposition entails, and nothing else.

2. Between t_1 and t_2 Jane learns P and nothing else. She updates by conditionalizing between those two times.
3. $cr_1(Q | P) = 2/3$.
4. $cr_3(Q | {\sim}P) = 1/2$.
5. $cr_3(P \supset Q) = cr_2(P \supset Q)$.
6. At t_3, Jane is certain of ${\sim}(P \,\&\, Q)$, anything that proposition entails, and nothing else.

(a) Completely specify Jane's credence distributions at t_2 and t_3.
(b) Create a hypothetical prior for Jane. In other words, specify a regular probabilistic distribution Pr_H over $\mathcal{L}$ such that cr_1 can be generated from Pr_H by conditionalizing on Jane's set of certainties at t_1; cr_2 is Pr_H conditionalized on Jane's certainties at t_2; and cr_3 is Pr_H conditionalized on Jane's t_3 certainties.
(c) Does Jane update by Conditionalization between t_2 and t_3? Explain how you know.
(d) The Hypothetical Priors Theorem says that *if* an agent always updates by conditionalizing, *then* her credences can be represented by a hypothetical prior distribution. Is the converse of this theorem true?

Problem 4.8. 𝕕𝕕𝕕 In this exercise we'll prove the Hypothetical Priors Theorem, by showing how to construct the needed Pr_H for any given series of credences.

Start by supposing we have a finite series of credence distributions $\{cr_1, cr_2, \ldots, cr_n\}$, each of which satisfies the probability axioms and Ratio Formula. Suppose also that each cr_i is related to cr_{i+1} as specified by Conditionalization. Then, for each cr_i, define E_i as a proposition logically equivalent to the conjunction of all the propositions receiving cr_i values of 1.[22]

(a) Let's start by focusing only on cr_1, the first credence distribution in the series. Describe how, given cr_1, to construct a regular probability distribution Pr_H such that cr_1 results from conditionalizing Pr_H on E_1. (It might help here to review the Ava example from page 111ff.)
(b) Now, using the cumulativity of conditionalization, explain why the fact that cr_1 results from conditionalizing Pr_H on E_1 guarantees that for every cr_i in the series, $cr_i(\cdot) = Pr_H(\cdot | E_i)$.

Problem 4.9. 𝕕𝕕𝕕 Suppose you have a finite partition $\{B_1, B_2, \ldots B_n\}$ of propositions from $\mathcal{L}$. Suppose also that between t_1 and t_2 you conditionalize on evidence equivalent to a disjunction of some of the Bs. Prove that for any A in $\mathcal{L}$ and any B_i such that $cr_2(B_i) > 0$,

$$cr_2(A \mid B_i) = cr_1(A \mid B_i)$$

Problem 4.10. ✐ Do you think only one set of ultimate epistemic standards is rationally permissible? Put another way: If two agents' series of credence distributions cannot be represented by the same hypothetical prior distribution, must at least one of them have assigned irrational credences at some point?

4.5 Further reading

INTRODUCTIONS AND OVERVIEWS

Ian Hacking (2001). *An Introduction to Probability and Inductive Logic*. Cambridge: Cambridge University Press

Chapter 15 works through many excellent examples of applying Bayes's Theorem to manage complex updates.

CLASSIC TEXTS

Rudolf Carnap (1947). On the Application of Inductive Logic. *Philosophy and Phenomenological Research* 8, pp. 133–48

Section 3 contains Carnap's original discussion of the Principle of Total Evidence.

EXTENDED DISCUSSION

Paul Teller (1973). Conditionalization and Observation. *Synthese* 26, pp. 218–58

Offers a number of arguments for the Conditionalization updating norm. (In Chapter 9 we'll discuss the Dutch Book argument for Conditionalization that Teller provides.)

Isaac Levi (1980). *The Enterprise of Knowledge*. Boston: The MIT Press

Though Levi's notation and terminology are somewhat different from mine, Chapter 4 thoroughly works through the mathematics of hypothetical priors.

Levi also discusses various historically important Bayesians' positions on how many distinct hypothetical priors are rationally permissible.

Christopher J.G. Meacham (2016). Ur-Priors, Conditionalization, and Ur-Prior Conditionalization. *Ergo* 3, pp. 444–92

Meacham considers a number of possible interpretations of hypothetical priors, and how they might be linked to an agent's credences at specific times by Conditionalization.

Notes

1. Remember that 1 is not a prime number, while 2 is!
2. A bit of reflection on Equation (4.2) will reveal that the constant by which we multiply in the second step of our process—the **normalization factor**—is always the reciprocal of the agent's initial unconditional credence in the evidence. In other words, the second step *divides* all nonzero state-description credences by $cr_i(E)$.
3. We can also now see an alternate explanation for steps (3.51) and (3.53) of Lewis's triviality proof from Section 3.3. The proposal assessed there is that for some conditional $\rightarrow$, the agent's conditional credence $cr(Z\,|\,Y)$ for any Y and Z in $\mathcal{L}$ equals her unconditional credence in $Y \rightarrow Z$. Whatever motivates that proposal, we should want the proposal to remain true even after the agent learns some information X. If the relevant values are going to match after conditionalization on X, it must be true before conditionalization that $cr(Y \rightarrow Z\,|\,X) = cr(Z\,|\,Y\,\&\,X)$, which is just Equation (3.59).
4. Thanks to Joel Velasco for discussion, and for the example.
5. For reasons we are now in a position to understand, the term "posterior" is sometimes used ambiguously in the Bayesian literature. I have defined "posterior" as an agent's conditional credence in the hypothesis given the evidence—$cr(H\,|\,E)$. If the agent updates by conditionalizing on E, this will equal her unconditional credence in the hypothesis after the update. The terms "prior" and "posterior" come from the fact that on an orthodox Bayesian position, those quantities pick out the agent's unconditional credences in the hypothesis before and after the update. But unorthodox Bayesians who prefer an alternative updating rule to Conditionalization nevertheless sometimes refer to an agent's post-update credence in a hypothesis as her "posterior". As I've defined the term, this is strictly speaking incorrect.
6. An acquaintance involved with neuroscientific research once claimed that when a prisoner in the American penal system comes up for parole, a particular kind of brain scan can predict with greater than 90% accuracy whether that prisoner will, if released, be sent back to jail within a specified period of time. He suggested that we use this brain scan in place of traditional parole board hearings, whose predictive accuracy is much lower. To get at some of my ethical concerns with this proposal, I asked why we don't just apply the brain scan to everyone in America, rather than waiting to see if a

person winds up in jail. He replied that the base rates make this impractical: While the recidivism rate among prisoners is fairly high, the percentage of Americans who wind up in jail is much lower, so the scan would generate far too many false positives if used on the general population.

7. In Section 2.2.3 I mentioned that Bayesians often work with an agent's set of doxastically possible worlds instead of the full set of logically possible worlds, understanding "mutually exclusive" and "tautology" in the Kolmogorov axioms in terms of this restricted doxastic set. The Regularity Principle concerns the *full* set of logically possible worlds—it forbids assigning credence 0 to any proposition that is true in at least one of them. So for the rest of this section, references to "contingent propositions", "tautologies", etc. should be read against that full logical set of possibilities.

8. What most of us call the "Regularity Principle" Dennis Lindley dubbed "Cromwell's Rule". He wrote, "A simple result that follows from Bayes' theorem is that it is inadvisable to attach probabilities of zero to uncertain events, for if the prior probability is zero so is the posterior, whatever be the data.... In other words, if a decision-maker thinks something cannot be true and interprets this to mean it has zero probability, he will never be influenced by *any* data, which is surely absurd. So leave a little probability for the moon being made of green cheese; it can be as small as 1 in a million, but have it there since otherwise an army of astronauts returning with samples of the said cheese will leave you unmoved.... As Oliver Cromwell told the Church of Scotland, 'I beseech you, in the bowels of Christ, think it possible you may be mistaken' " (Lindley 1985, p. 104, emphasis in original). (Thanks to Patrick Cronin for bringing this to my attention and to Wikipedia for the reference.)

9. Throughout this section I identify credence 1 with absolute certainty in a proposition and credence 0 with ruling that proposition out. This becomes more complicated when we consider events with infinitely many possible outcomes; the relevant complications will be addressed in Chapter 5.

10. Observation selection effects pop up all over the place in real life—perhaps you think the refrigerator light is *always* on because it's on whenever you open the door to look. Here's my favorite example: During World War II, the American military showed mathematician Abraham Wald data indicating that planes returning from dogfights had more bullet holes in the fuselage than in the engine. The military was considering shifting armor from the engine to the fuselage. Wald recommended exactly the opposite, on grounds that it was the *returning* planes that had holes in the fuselage but not the engines (Wainer 2011, recounted in Ellenberg 2014, pp. 12–3).

11. A similar situation occurs in Bradley (2010): Colin Howson argued that a so-called "Thomason case" provides a counterexample to Conditionalization. Bradley replies that if we analyze the agent's *total* evidence in the case—including evidence about how he came to have his evidence—the supposed counterexample disappears.

12. You may have noticed that in the Monty Hall Problem, accounting for the agent's total relevant evidence required us to move from a coarser-grained partition of her doxastic possibilities (Prize behind door A/B/C) to a finer-grained partition (Prize behind A & Monty reveals B, Prize behind A & Monty reveals C, etc.). Whether a conditionalization yields the right results often depends on the richness of the language in which the agent represents her doxastic possibilities; a language without enough detail may miss aspects

of her total relevant evidence. For more on selecting an appropriately detailed language for updating, and some formal results on how one can know when one's language is rich enough, see Titelbaum (2013a, Ch. 8).

13. I learned of Lewis's "superbaby" talk from Alan Hájek. Susan Vineberg suggested to me that Lewis's inspiration for the term may have been I.J. Good's (1968) discussion of "an infinitely intelligent newborn baby having built-in neural circuits enabling him to deal with formal logic, English syntax, and subjective probability"—a discussion to which we shall return in Chapter 6.

14. The earliest sustained usage I can find of the phrase "epistemic standards" occurs in Foley (1987, p. 33ff.). While Foley may have coined the *name*, he wasn't the first to discuss the *notion* I'm calling "ultimate epistemic standards"—Levi (1980) has a similar notion of "confirmational commitments", and in Chapter 6 we'll see Carnap proposing a similar thing much earlier. Levi's discussion is particularly important because it lays out the mathematical formalism for epistemic standards I'm about to present.

While we're discussing terminology: Given the choice, I'd prefer the term "evidential standards" to "epistemic standards", because the former emphasizes the standards' taking evidence as their inputs, while the latter introduces an unwanted association with knowledge. Yet the "epistemic standards" terminology is so entrenched at this point that I've given up fighting against it. (Thanks to Laura Callahan for the pointer to Foley.)

15. Compare: When we speak admiringly of someone's "high moral standards", I don't think we're necessarily attributing the possession of systematic moral principles. This way of talking seems as compatible with virtue ethics as any other moral approach.

16. Catrin Campbell-Moore pointed out to me that the Hypothetical Priors Theorem assumes that a contingent proposition receives credence 1 at t_i only if it is entailed by E_i. If this assumption fails, we will still be able find at least one hypothetical prior Pr_H for the series of credence distributions, but no *regular* hypothetical prior will be available.

The assumption reflects the idea that the only thing that can justify certainty in a contingent proposition is empirical evidence. It's interesting to think about cases in which this assumption might fail. It might be that particular propositions, despite being logically contingent, nevertheless merit rational a priori certainty. Or another way the assumption could fail is if credence 1 doesn't always represent certainty. We'll see examples where this might occur in Section 5.4.

17. This is why defining hypothetical priors as regular does not commit us to the Regularity Principle as a constraint on rational agents' credences. We make hypothetical priors regular so they will remain independent of contingent evidence, and therefore capable of representing extra-evidential influences on an agent's attitudes. But since hypothetical priors aren't credences, this doesn't say anything about whether credences should be regular or not. Moreover, the Hypothetical Priors Theorem applies only to agents who update by Conditionalization, while Conditionalization conflicts with the Regularity Principle.

18. In the Bayesian literature, the terms "initial prior", "ur-prior", and "hypothetical prior" are often used interchangeably. To me, the former two connote that the prior was assigned by the agent at some early time. So I've selected "hypothetical prior" to emphasize the use of a mathematical representation that does not correspond to any credences the agent ever actually assigned. Unfortunately, the term "hypothetical prior"

has also been used for a very specific notion within the literature on the Problem of Old Evidence (as in Bartha and Hitchcock 1999, p. S349). Here I simply note the distinction between that usage and the one I intend; I'll explain the alternate usage in Section 12.1.3.

19. This is a version of a problem from Julia Staffel.

20. This problem was inspired by a problem of Sarah Moss's.

21. This story is adapted from Hart and Titelbaum (2015).

22. We define E_i this way for reasons discussed in note 16.

5

Further Rational Constraints

The previous three chapters have discussed five core normative Bayesian rules: Kolmogorov's three probability axioms, the Ratio Formula, and Conditionalization. Bayesians offer these rules as necessary conditions for an agent's credences to be rational. We have not discussed whether these five rules are jointly sufficient for rational credence.

Agents can satisfy the core rules and still have wildly divergent attitudes. Suppose 1,000 balls have been drawn from an urn and every one of them has been black. In light of this evidence, I might be highly confident that the next ball drawn will be black. But I might also have a friend Mina, whose credences satisfy all the rational constraints we have considered so far, yet who nevertheless responds to the same evidence by being 99% confident that the next ball will be white. Similarly, if you tell me you rolled a fair die but don't say how the roll came out, I might assign credence 1/6 that it came up six. Mina, however, could be 5/6 confident of that proposition, without violating the core Bayesian rules in any way.

If we think Mina's credences in these examples are irrational, we need to identify additional rational requirements beyond the Bayesian core that rule them out. We have already seen one potential requirement that goes beyond the core: the Regularity Principle (Section 4.2) prohibits assigning credence 0 to logically contingent propositions. What other requirements on rational credence might there be? When all the requirements are put together, are they strong enough to dictate a single rationally permissible credence distribution for each possible body of total evidence?

The answer to this last question is sometimes taken to separate Subjective from Objective Bayesians. Unfortunately, "Objective/Subjective Bayesian" terminology is used ambiguously, so this chapter begins by clarifying two different ways in which those terms are used. In the course of doing so we'll discuss various interpretations of probability, including frequency and propensity views.

Then we will consider a number of additional rational credence constraints proposed in the Bayesian literature. We'll begin with synchronic constraints: the Principal Principle (relating credences to chances); the Reflection Principle

Fundamentals of Bayesian Epistemology 1: Introducing Credences. Michael G. Titelbaum, Oxford University Press.

(concerning one's current credences about one's future credences); principles for deferring to experts; indifference principles (for distributing credences in the absence of evidence); and principles for distributing credences over infinitely many possibilities. Finally, we will turn to Jeffrey Conditionalization, a diachronic updating principle proposed as a generalization of standard Conditionalization.

Most of these constraints are usually offered as *supplements* to the five core Bayesian rules we've seen already. You may not have noticed, but in discussing Conditionalization and drawing out its consequences, we assumed throughout that the updating agents satisfied the probability axioms and Ratio Formula. Similarly, most of the principles we will discuss in this chapter build upon the five core rules, and only have their intended effects if those five rules are assumed in the background. Jeffrey Conditionalization, on the other hand, is sometimes proposed as a *substitute* for Conditionalization—though it still assumes the other four, synchronic core rules.

5.1 Subjective and Objective Bayesianism

When a weather forecaster comes on television and says, "The probability of snow tomorrow is 30%," what does that mean? What exactly has this forecaster communicated to her audience? Such questions have been asked throughout the history of mathematical probability theory; in the twentieth century, rival answers became known as **intepretations of probability**. There is an excellent literature devoted to this topic and its history (see the Further Reading of this chapter for recommendations), so I don't intend to let it take over this book. But for our purposes it's important to touch on some of the main interpretations, and at least mention some of their advantages and disadvantages.

5.1.1 Frequencies and propensities

The earliest European practitioners of mathematical probability theory applied what we now call the **classical interpretation** of probability. This interpretation, championed most famously by Pierre-Simon Laplace, calculates the probability of a proposition by counting up the number of possible event outcomes consistent with that proposition, then dividing by the total number of outcomes possible. For example, if I roll a six-sided die, there are six possible outcomes, and three of them are consistent with the proposition that the die came up even. So the classical probability of even is 1/2. (This is almost certainly the kind of probability you first encountered in school.)

Laplace advocated this procedure for any situation in which "nothing leads us to believe that one of [the outcomes] will happen rather than the others" (Laplace 1814/1995, p. 3). Applying what Jacob Bernoulli (1713) had earlier called the "principle of insufficient reason", Laplace declared that in such cases we should view the outcomes as "equally possible", and calculate the probabilities as described above.

The notion of "equally possible" at the crux of this approach clearly needs more philosophical elucidation. But even setting that aside, the classical interpretation leaves us adrift the moment someone learns to shave a die. With the shape of the die changed, our interpretation of probability needs to allow the possibility that some faces are more probable than others. For instance, it might now be 20% probable that you will roll a six. While Laplace recognized and discussed such cases, it's unclear how his view can interpret the probabilities involved. There are no longer possible outcomes of the roll that can be tallied up and put into a ratio equaling 20%.

So suppose a shady confederate offers to sell you a shaved die with "a 20% probability of landing six". How might she explain—or back up—that probability claim? Well, if an event has a 20% probability of producing a certain outcome, we expect that were the event repeated it would produce that type of outcome roughly 20% of the time. The **frequency theory** of probability uses this fact to analyze "probability" talk. On this interpretation, when your confederate claims the die has a 20% probability of landing six on a given roll, she *means* that repeated rolls of the die will produce a six about 20% of the time. According to the frequency theory, the probability is x that event A will have outcome B just in case proportion x of events like A have outcomes like B.[1] The frequency theory originated in work by Robert Leslie Ellis (1849) and John Venn (1866), then was famously developed by the logical positivist Richard von Mises (1928/1957).

The frequency theory has a number of problems; I will mention only a few.[2] Suppose my sixteen-year-old daughter asks for the keys to my car; I wonder what the probability is that she will get into an accident should I give her the keys. According to the frequency theory, the probability that the event of my giving her the keys will have the outcome of an accident is determined by how frequently this type of event leads to accidents. But what type of event is it? Is the frequency in question how often people who go driving get into accidents? How often sixteen-year-olds get into accidents? How often sixteen-year-olds with the courage to ask their fathers for the keys get into accidents? How often my daughter gets into accidents? Presumably these frequencies will differ—which one is *the* probability of an accident should I give my daughter the keys right now?

Any event can be subsumed under many types, and the frequency theory leaves it unclear which event-types determine probability values. Event types are sometimes known as reference classes, so this is the **reference class problem**. In response, one might suggest that outcomes have frequencies—and therefore probabilities—only *relative* to the specification of a particular reference class (either implicitly or explicitly). But it seems we can meaningfully inquire about the probabilities of particular event outcomes (or of propositions simpliciter) without specifying a reference class. I need to decide whether to give the keys to my daughter; I want to know how probable it is that she will crash. That probability doesn't seem relative to any particular reference class. Or if it is (covertly) relative to some reference class, which reference class does the job?

Frequency information about specific event-types seems more relevant to determining probabilities than information about general types. (The probability that my daughter will get into an accident on this occasion seems much closer to *her* frequency of accidents than to the accident frequency of drivers in general.) Perhaps probabilities are frequencies in the maximally specific reference class? But the *maximally* specific reference class containing a particular event contains only that individual event. The frequency with which my daughter gets into an accident when I give her my keys *on this occasion* is either 0 or 1—but we often think probabilities for such events have nonextreme values.

This brings us to another problem for frequency theories. Suppose I have a penny, and think that if I flip it, the probability that the flip will come out heads is 1/2. Let's just grant *arguendo* that the correct reference class for this event is penny flips. According to the frequency theory, the probability that this flip will come up heads is the fraction of all penny flips that ever occur which come up heads. Yet while I'd be willing to bet that fraction is *close* to 1/2, I'd be willing to bet even more that the fraction is not *exactly* 1/2. (For one thing, the number of penny flips that will ever occur in the history of the universe might be an odd number!) For any finite run of trials of a particular event-type, it seems perfectly coherent to imagine—indeed, to *expect*—that a particular outcome will occur with a frequency not precisely equal to that outcome's probability. Yet if the frequency theory is correct, this is *conceptually impossible* when the run in question encompasses every event trial that will ever occur.

One might respond that the probability of heads on the flip of a penny is not the frequency with which penny flips *actually* come up heads over the finite history of our universe; instead, it's the frequency *in the limit*—were pennies to continue being flipped forever. This gives us **hypothetical frequency theory**, on which the probability of an outcome is the frequency it would approach were the event repeated indefinitely. Yet this move undermines one of the original appeals of the frequency approach: its empiricism. The proportion of event repetitions that produce a particular outcome in the actual world is the sort of thing that could be *observed* (at least in principle)—providing a sound empirical base for otherwise-mysterious "probability" talk. Empirically grounding hypothetical frequencies is a much more difficult task.

Moreover, there seem to be events that *couldn't* possibly be repeated many many times, and even events that couldn't be repeated once. Before the Large Hadron Collider was switched on, physicists were asked for the probability that doing so would destroy the Earth. Were that to have happened, switching on the Large Hadron Collider would not have been a repeatable event. Scientists also sometimes discuss the probability that our universe began with a Big Bang; arguably, that's not an event that will happen over and over or even *could* happen over and over. So it's difficult to understand talk about how frequently the universe would begin with a Bang were the number of times the universe started increased toward the limit. This problem of assigning meaningful nonextreme probabilities to individual, perhaps non-repeatable events is called the **problem of the single case**.

The frequentist still has moves available. Faced with a single event that's non-repeatable in the actual world, she might ask what proportion of times that event produces a particular outcome across *other* possible worlds.[3] But now the prospects for analyzing "probability" talk in empirically observable terms have grown fairly dim.

An alternate interpretation of probability admits that probabilities are related to frequencies, but draws our attention to the features that *cause* particular outcomes to appear with the frequencies that they do. What is it about a penny that makes it come up heads about half the time? Presumably something about its physical attributes, the symmetries with which it interacts with surrounding air as it flips, etc. These traits lend the penny a certain tendency to come up heads, and an equal tendency to come up tails. This quantifiable disposition—or **propensity**—would generate certain frequencies were a long run of trials to be staged. But the propensity is also at work in each individual flip, whether that flip is ever repeated or could ever be repeated.

A non-repeatable experimental setup may possess a nonextreme propensity to generate a particular outcome.

While an early propensity theory appeared in the work of Charles Sanders Peirce (1910/1932), propensity's most famous champion was Karl Popper (1957). Popper was especially motivated by developments in quantum mechanics. In quantum theory the Born rule calculates probabilities of experimental outcomes from a particular quantity (the amplitude of the wave-function) with independent significance in the theory's dynamics. Moreover, this quantity can be determined for a particular experimental setup even if that setup is never to be repeated (or *couldn't* be repeated) again. This gives propensities a respectable place within an empirically established scientific theory. Propensities may also figure in such theories as statistical mechanics and population genetics.

Yet even if there are propensities in the world, it seems difficult to interpret *all* probabilities as propensities. Suppose we're discussing the likelihood that a particular outcome occurs given that a quantum experiment is set up in a particular fashion. This is a conditional probability, and it has a natural interpretation in terms of physical propensities: the experimental setup described in the condition of the conditional probability has a particular causal tendency to produce the outcome. But where there is a likelihood, probability mathematics suggests there will also be a posterior—if there's a probability of outcome given setup, there should also be a probability of setup given outcome. Yet the latter hardly makes sense as a physical propensity—does an experimental outcome have a quantifiable causal tendency to produce the particular experimental setup from which it results?[4]

Some philosophers—especially those of a Humean bent—are also suspicious of the metaphysics of propensities. From their point of view, causes are objectionable enough; even worse to admit propensities that seem to be a kind of graded causation. Nowadays most philosophers of science agree that we need some notion of physical probability that applies to the single case. Call this notion **objective chance**. But whether objective chances are best understood via propensity theory, a "best systems" analysis (Lewis 1994), or some other approach is a hotly contested matter.

Finally, whatever objective chances turn out to be, they are governed by the physical laws of our world. That means there can be no objective chance that the physical laws are one way or another. (What set of laws beyond the physical might determine such chances?) Yet it seems physicists can meaningfully discuss the probability that the physical laws of the universe will turn out to be such-and-such. While the notion of objective chance makes sense of *some* of our "probability" talk, it nevertheless seems to leave a remainder.

5.1.2 Two distinctions in Bayesianism

So what *are* physicists talking about when they discuss the probability that the physical laws of the universe are one way rather than another? Perhaps they are expressing their degrees of confidence in alternative physical hypotheses. Perhaps there are no probabilities out in the world, independent of us, about which our opinions change as we gain evidence. Instead, it may be that facts in the world are simply true or false, probability-free, and "probability" talk records our changing confidences in those facts in the face of changing evidence.

Bayesian theories are often characterized as "Subjective" or "Objective", but this terminology can be used to draw two different distinctions. One of them concerns the interpretation of "probability" talk. On this distinction—which I'll call the **semantic distinction**—Subjective Bayesians adopt the position I proposed in the previous paragraph. For them, "probability" talk expresses or reports the degrees of confidence of the individuals doing the talking, or perhaps of communities to which they belong. Objective Bayesians, on the other hand, interpret "probability" assertions as having truth-conditions independent of the attitudes of particular agents or groups of agents.[5] In the twentieth century, talk of "Objective" and "Subjective" Bayesianism was usually used to draw this semantic distinction.[6]

More recently the "Subjective Bayesian/Objective Bayesian" terminology has been used to draw a different distinction, which I will call the **normative distinction**. However we interpret the meaning of "probability" talk, we can grant that agents assign different degrees of confidence to different propositions (or, more weakly, that it is at least useful to model agents as if they do). Once we grant that credences exist and are subject to rational constraints, we may inquire about the strength of those constraints.

On one end of the spectrum, Objective Bayesians (in the normative sense) endorse what Richard Feldman (2007) and Roger White (2005) have called the

Uniqueness Thesis: Given any proposition and body of total evidence, there is exactly one attitude it is rationally permissible for agents with that body of total evidence to adopt toward that proposition.

Assuming the attitudes in question are degrees of belief, the Uniqueness Thesis says that given any evidential situation, there's exactly one credence that any agent is rationally permitted to adopt toward a given proposition in that situation. The Uniqueness Thesis entails **evidentialism**, according to which the attitudes rationally permissible for an agent supervene on her evidence.

Suppose we have two agents with identical total evidence who adopt different credences toward some propositions. Because Objective Bayesians (in the normative sense) endorse the Uniqueness Thesis, they will maintain that at least one of these agents is responding to her evidence irrationally. In most real-life situations, different agents have different bodies of total evidence— and even different bodies of *relevant* evidence—so many discrepancies in their attitudes can be chalked up to evidential differences. But we have stipulated in this case that the agents have identical evidence, so whatever causes the differences in their attitudes, it can't be the contents of their evidence. In Section 4.3 we identified the extra-evidential factors that determine an agent's attitudes in light of her total evidence as her "ultimate epistemic standards". These epistemic standards might reflect pragmatic influences, a predilection for hypotheses with certain features, a tendency toward mistrust or skepticism, etc.

The Hypothetical Priors Theorem tells us that whenever an agent's credence distributions over time satisfy the probability axioms, Ratio Formula, and Conditionalization, her epistemic standards can be represented by a hypothetical prior distribution. This regular, probabilistic distribution stays constant as the agent gains evidence over time. Yet we can always recover the agent's credence distribution at a given time by conditionalizing her hypothetical prior on her total evidence at that time.

The core Bayesian rules (probability axioms, Ratio Formula, Conditionalization) leave a wide variety of hypothetical priors available. Assuming they satisfy the core rules, our two agents who assign different credences in response to the same total evidence must have different hypothetical priors. According to the Objective Bayesian (in the normative sense), any time such a situation arises at least one of the agents must be violating rational requirements. Thus the Objective Bayesian thinks there is exactly one set of rationally permissible hypothetical priors—one set of correct epistemic standards embodying rational agents' common responses to evidence.

How might the unique rational hypothetical prior be generated, and how might we justify the claim that it is uniquely correct? Our *ongoing* epistemic standards for responding to new pieces of empirical evidence are often informed by other pieces of evidence we gained in the past. I react to a fire alarm in a particular way because I've experienced such alarms before; one piece of evidence helps determine how we interpret the next. But *ultimate* epistemic standards—the ones represented by our hypothetical priors—dictate responses to our *total* evidence, and so must be rationally antecedent to all of our evidence. If we are to select and justify a unique set of ultimate epistemic standards, we must do so a priori.

Extending a tradition that dated back to Bolzano (1837/1973) and perhaps even Leibniz,[7] Keynes (1921) and Carnap (1950) argued that just as there are objective facts about which propositions are logically *entailed* by a given body of evidence, there are objective logical facts about the degree to which a body of evidence probabilifies a particular proposition. Carnap went on to offer a mathematical algorithm for calculating the unique logical hypothetical priors from which these facts could be determined; we will discuss that algorithm in Chapter 6. (The **logical interpretation** of probability holds that an agent's "probability" talk concerns logical probabilities relative to her current total evidence.)[8] Many recent theorists, while backing away from Keynes's and Carnap's position that these values are *logical*, nevertheless embrace the idea of **evidential probabilities** reflecting the degree to which a proposition is probabilified by a given body of evidence. If you think that rationality requires an agent to assign credences equal to the unique, true evidential probabilities on her current total evidence, you have an Objective Bayesian view in the normative sense.[9]

At the other end of the spectrum from Objective Bayesians (in the normative sense) are theorists who hold that the probability axioms and Ratio Formula are the only rational constraints on hypothetical priors.[10] The literature often defines "Subjective Bayesians" as people who hold this view. But that terminology leaves no way to describe theorists in the middle of the spectrum—the vast majority of Bayesian epistemologists who believe in rational constraints on hypothetical priors that go beyond the core rules but are insufficient to narrow things down to a single permissible standard. I will use the term "Subjective Bayesian" (in the normative sense) to refer to anyone who thinks more than one hypothetical prior is rationally permissible. I will call people who think the Ratio Formula and probability axioms are the only rational constraints on hypothetical priors "extreme Subjective Bayesians".

Subjective Bayesians allow for what White calls **permissive cases**: situations in which two agents reach different conclusions on the basis of the same total evidence without either party's making a rational mistake. This is because each agent interprets the evidence according to different (yet rationally acceptable) epistemic standards, which allow them to draw different conclusions.

I have distinguished the semantic and normative Objective/Subjective Bayesian distinctions because they can cross-cut one another. Historically, Ramsey (1931) and de Finetti (1931/1989) reacted to Keynes's Objective Bayesianism with groundbreaking theories that were Subjective in both the semantic and normative senses. But one could be a Subjective Bayesian in the semantic sense—taking agents' "probability" talk to express their own current credences—while maintaining that strictly speaking only one credence distribution is rationally permitted in each situation (thereby adhering to Objective Bayesianism in the normative sense). Going in the other direction, one could admit the existence of degrees of belief while holding that they're not what "probability" talk concerns. This would give an Objective Bayesian semantic view that combined with either Subjective or Objective Bayesianism in the normative sense. Finally, probability semantics need not be monolithic; many Bayesians now hold that some "probability" assertions in everyday life express credences, others report objective chances, and still others indicate what would be reasonable to believe given one's evidence.[11]

Regardless of her position on the semantics, any Bayesian who isn't an extreme Subjective Bayesian in the normative sense will concede that there are rational constraints on agents' hypothetical priors beyond the probability axioms and Ratio Formula. The rest of this chapter investigates what some of those additional constraints might be. I should note at the outset, though, that the more powerful and widely applicable these constraints get, the more they seem to be beset by problems. Many Subjective Bayesians (in the normative sense) would be happy to adopt an Objective position, if only they could see past the numerous shortcomings of the principles Objective Bayesians propose to generate unique hypothetical priors. Richard Jeffrey characterized his Subjective Bayesian position as follows:

> As a practical matter, I think one can give necessary conditions for reasonableness of a set of partial beliefs that go beyond mere [probabilistic] coherence—in special cases. The result is a patch-work quilt, where the patches have frayed edges, and there are large gaps where we lack patches altogether. It is not the sort of seamless garment philosophers like to wear; but (we ragged pragmatists say), the philosophers are naked! Indeed we have no proof that no more elegant garb than our rags is available, or ever will be, but we haven't seen any, yet, as far as we know. We will be the first to snatch it off the racks, when the shipments come in. But perhaps they never will. Anyway, for the time being, we are dressed in rags, tied neatly at the waist with a beautiful cord—probabilistic coherence. (It is the only cord that visibly distinguishes us from the benighted masses.)   (1970, p. 169)

5.2 Deference principles

5.2.1 The Principal Principle

Bayesian epistemology concerns agents' degrees of belief. Yet most contemporary Bayesian epistemologists also believe that the world contains objective chances of some sort—physical probabilities that particular events will produce particular outcomes. This raises the question of how subjective credences and objective chances should relate. One obvious response is a principle of direct inference: roughly, rational agents set their credences in line with what they know of the chances. If you're certain a die is fair (has an equal objective chance of landing on each of its faces), you should assign equal credence to each possible roll outcome.

While direct inference principles have a long history, the most famous such principle relating credence and chance is David Lewis's (1980) Principal Principle. The Principal Principle's most straightforward consequence is that if you are certain an event has objective chance x of producing a particular outcome, and you have no other information about that event, then your credence that the outcome will occur should be x. For many Bayesian purposes this is all one needs to know about the Principal Principle. But in fact the Principle is a more finely honed instrument, because Lewis wanted it to deal with complications like the following: (1) What if you're uncertain about the objective chance of the outcome? (2) What if the outcome's chance changes over time? (3) What if you have additional information about the event besides what you know of the chances? The rest of this section explains how the Principal Principle deals with those eventualities. If you're not interested in the details, feel free to skip to Section 5.2.2.

So: Suppose it is now 1 p.m. on a Monday. I tell you that over the weekend I found a coin from a foreign country that is somewhat irregular in shape. Despite being foreign, one side of the coin is clearly the "Heads" side and the other is "Tails". I also tell you that I flipped the foreign coin today at noon.

Let H be the proposition that the noon coin flip landed heads. Consider each of the propositions below one at a time, and decide what your credence in H would be if that proposition was *all* you knew about the coin in addition to the information in the previous paragraph:

E_1: After discovering the coin I spent a good part of my weekend flipping it, and out of my 100 weekend flips sixty-four came up heads.

E_2:   The coin was produced in a factory that advertises its coins as fair, but also has a side business generating black-market coins biased toward tails.

E_3:   The coin is fair (has a 1/2 chance of landing heads).

E_4:   Your friend Amir was with me at noon when I flipped the coin, and he told you it came up heads.

Hopefully it's fairly clear how to respond to each of these pieces of evidence, taken singly. For instance, in light of the frequency information in E_1, it seems rational to have a credence in H somewhere around 0.64. We might debate whether precisely 0.64 is required,[12] but certainly a credence in H of 0.01 (assuming E_1 is your *only* evidence about the coin) seems unreasonable.

This point generalizes to a rational principle that whenever one's evidence includes the frequency with which events of type A have produced outcomes of type B, one should set one's credence that the next A-event will produce a B-outcome equal to (or at least in the vicinity of) that frequency.[13] While some version of this principle ought to be right, working out the specifics creates problems like those faced by the frequency interpretation of probability. For instance, we have a reference class problem: Suppose my evidence includes accident frequency data for drivers in general, for sixteen-year-old drivers in general, and for my sixteen-year-old daughter in particular. Which value should I use to set my credence that my daughter will get in a car accident tonight? The more specific data seems more relevant, but the more general data reflects a larger sample.[14]

There are statistical tools available for dealing with these problems, some of which we will discuss in Chapter 13. But for now let's focus on a different question about frequency data: *Why* do we use known flip outcomes to predict the outcome of unobserved flips? Perhaps because known outcomes indicate something about the physical properties of the coin itself; they help us figure out its objective chance of coming up heads. Known flip data influence our unknown flip predictions because they make us think our coin has a particular chance profile. In this case, frequency data influences predictions *by way of* our opinions about objective chances.

This relationship between frequency and chance is revealed when we combine pieces of evidence listed above. We've already said that if your only evidence about the coin is E_1—it came up heads on sixty-four of 100 known tosses—then your credence that the noon toss (of uncertain outcome) came up heads should be around 0.64. On the other hand, if your only evidence is E_3, that the coin is fair, then I hope it's plausible that your credence in H should be 0.5. But what if you're already certain of E_3, and then learn E_1? In that case your credence in heads should still be 0.5.

Keep in mind we're imagining you're *certain* that the coin is fair before you learn the frequency data; we're not concerning ourselves with the possibility that, say, learning about the frequencies makes you suspicious of the source from which you learned that the coin is fair. If it's a fixed, unquestionable truth for you that the coin is fair, then learning that it came up sixty-four heads on 100 flips will not change your credence in heads. If *all* you had was the frequency information, that would support a different hypothesis about the chances. But it's not as if sixty-four heads on 100 flips is *inconsistent* with the coin's being fair—a fair coin usually won't come up heads on exactly half the flips in a given sample. So once you're already certain of heads, the frequency information becomes redundant, irrelevant to your opinions about unknown flips. Frequencies help you learn about chances, so if you are already certain of the chances there's nothing more for frequency information to do.

David Lewis called information that can change your credences about an event only *by way of* changing your opinions about its chances **admissible** information. His main insight about admissible information was that when the chance values for an event have already been established, admissible information becomes irrelevant to a rational agent's opinions about the outcome.

Here's another example: Suppose your only evidence about the noon flip outcome is E_2, that the coin was produced in a factory that advertises its coins as fair but has a side business in tails-biased coins. Given only this information your credence in H should be somewhere below 0.5. (Exactly how far below depends on how extensive you estimate the side business to be.) On the other hand, suppose you learn E_2 after already learning E_3, that this particular coin is fair. In that context, E_2 becomes irrelevant, at least with respect to predicting flips of this coin. E_2 is relevant in isolation because it informs you about the chances associated with the coin. But once you're certain that the coin is fair, information E_2 only teaches you that you happened to get lucky not to have a black-market coin; it doesn't do anything to push your credence in H away from 0.5. E_2 is admissible information.

Contrast that with E_4, your friend Amir's report that he observed the flip landing heads. Assuming you trust Amir, E_4 should make you highly confident in H. And this should be true even if you already possess information E_3 that the coin is fair. Notice that E_3 and E_4 are consistent; the coin's being fair is consistent with its having landed heads on this particular flip, and with Amir's reporting that outcome. But E_4 trumps the chance information; it moves your credence in heads away from where it would be (0.5) if you knew only E_3. Information about this particular flip's outcome does not change your credences about the flip *by way of* influencing your opinions about the chances. You still think the coin is fair, and was fair at the time it was flipped. You just know now that the fair coin happened to come up heads on this occasion. Information about this flip's outcome is inadmissible with respect to H.

Lewis expressed his insight about the irrelevance of admissible information in his famous chance-credence principle, the

Principal Principle: Let Pr_H be any reasonable initial credence function. Let t_i be any time. Let x be any real number in the unit interval. Let $Ch_i(A) = x$ be the proposition that the chance, at time t_i, of A's holding equals x. Let E be any proposition compatible with $Ch_i(A) = x$ that is admissible at time t_i. Then

$$Pr_H(A \mid Ch_i(A) = x \,\&\, E) = x$$

(I have copied this principle verbatim from Lewis 1980, p. 266, though I have altered Lewis's notation to match our own.) There's a lot to unpack in the Principal Principle, so we'll take it one step at a time. First, Lewis's "reasonable initial credence function" sounds a lot like an initial prior distribution. Yet we saw in Section 4.3 that the notion of an initial prior is problematic, and there are passages in Lewis that make it sound more like he's talking about a hypothetical prior.[15] So I will interpret the "reasonable initial credence function" as your hypothetical prior distribution, and designate it with our notation "Pr_H".

The Principal Principle is proposed as a rational constraint on hypothetical priors, one that goes beyond the probability axioms and Ratio Formula. Why frame the Principal Principle around hypothetical priors, instead of focusing on the credences of rational agents at particular times? One advantage of the hypothetical-priors approach is that it makes the total evidence at work explicit, and therefore easy to reference in the principle. Recall from Section 4.3 that a hypothetical prior is a probabilistic, regular distribution containing no contingent evidence. A rational agent is associated with a particular hypothetical prior, in the sense that if you conditionalize that hypothetical prior on the agent's total evidence at any given time, you get the agent's credence distribution at that time.

In the Principal Principle we imagine that a real-life agent is considering some proposition A about the outcome of a chance event. She has

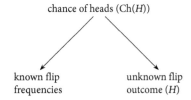

Figure 5.1 Chances screen off frequencies

some information about the chance of A, $Ch_i(A) = x$, and then some further evidence E. So her total evidence is $Ch_i(A) = x$ & E, and by the definition of a hypothetical prior her credence in A equals $Pr_H(A \mid Ch_i(A) = x$ & $E)$. Lewis claims that as long as E is both admissible for A, and is compatible (which we can take to mean "logically consistent") with $Ch_i(A) = x$, E should make no difference to the agent's credence in A. In other words, as long as E is admissible and compatible, the agent should be just as confident in A as she would be if all she knew were $Ch_i(A) = x$. That is, her credence in A should be x.

Return to our example about the noon coin flip, and the relationship between chance and frequency information. Suppose that at 1 p.m. your total evidence about the flip outcome consists of E_1 and E_3. E_3, the chance information, says that $Ch(H) = 0.5$. E_1, the frequency information, comprises the rest of your total evidence, which will play the role of E in the Principal Principle. Because this additional evidence is both consistent with $Ch(H) = 0.5$ and admissible for H, the Principal Principle says your 1 p.m. credence in H should be 0.5. Which is exactly the result we came to before.

We can gain further insight into this result by connecting it to our earlier (Section 3.2.4) discussion of causation and screening off. Figure 5.1 illustrates the causal relationships in the coin example between chances, frequencies, and unknown results. The coin's physical structure, associated with its objective chances, causally influenced the frequency with which it came up heads in the previous trials. The coin's physical makeup also affects the outcome of the unknown flip. Thus previous frequency information is relevant to the unknown flip, but only by way of the chances.[16] We saw in Section 3.2.4 that when this kind of causal fork structure obtains, the common cause screens its effects off from each other.[17] Conditional on the chances, frequency information becomes irrelevant to flip predictions. That is,

$$Pr_H(H \mid Ch(H) = 0.5 \text{ \& } E) = Pr_H(H \mid Ch(H) = 0.5) \qquad (5.1)$$

and intuitively the expression on the right should equal 0.5.

A similar analysis applies if your total evidence about the coin flip contains only $Ch(H) = 0.5$ and E_2, the evidence about the coin factory. This time

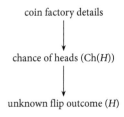

coin factory details

chance of heads (Ch(*H*))

unknown flip outcome (*H*)

Figure 5.2 Chance in a causal chain

our structure is a causal chain, as depicted in Figure 5.2. The situation in the coin factory causally affects the chance profile of the coin, which in turn causally affects the unknown flip outcome. Thus the coin factory information affects opinions about H by way of the chances, and if the chances are already determined then factory information becomes irrelevant. Letting the factory information play the role of E in the Principal Principle, the chances screen off E from H and we have the relation in Equation (5.1).

Finally, information E_4, your friend Amir's report, is not admissible information about H. E_4 affects your opinions about H, but not by way of affecting your opinions about the chances. The Principal Principle applies only when E, the information possessed in addition to the chances, is admissible. Since E_4 is inadmissible, the Principal Principle supplies no guidance about setting your credences in light of it.

There are still a few details in the principle to unpack. For instance, the chance expression $Ch_i(A)$ is indexed to a time t_i. That's because the chance that a particular proposition will obtain can change as time goes on. For instance, suppose that at 11am our foreign coin was fair, but at 11:30 I stuck a particularly large, non-aerodynamic wad of chewing gum to one of its sides. In that case, the proposition H that the coin comes up heads at noon would have a chance of 0.5 at 11am but might have a different chance after 11:30. The physical details of an experimental setup determine its chances, so as physical conditions change chances may change as well.[18]

Finally, the Principal Principle's formulation in terms of conditional credences allows us to apply it even when an agent doesn't have full information about the chances. Suppose your total evidence about the outcome A of some chance event is E. E influences your credences in A by way of informing you about A's chances (so E is admissible), but E does not tell you what the chances are exactly. Instead, E tells you that the chance of A (at some specific time, which I'll suppress for the duration of this example) is either 0.7 or 0.4. E also supplies you with a favorite among these two chance hypotheses: it sets your credence that 0.7 is the true chance at 2/3, and your credence that 0.4 is the true chance at 1/3.

How can we analyze this situation using the Principal Principle? Since your total evidence is E, the definition of a hypothetical prior distribution tells us that your current credences cr should be related to your hypothetical prior Pr_H as follows:

$$cr(A) = Pr_H(A \mid E) \qquad (5.2)$$

This value is not dictated directly by the Principal Principle. However, the Principal Principle does set

$$Pr_H(A \mid Ch(A) = 0.7 \ \& \ E) = 0.7 \qquad (5.3)$$

because we stipulated that E is admissible. Similarly, the Principal Principle sets

$$Pr_H(A \mid Ch(A) = 0.4 \ \& \ E) = 0.4 \qquad (5.4)$$

Since E narrows the possibilities down to two mutually exclusive chance hypotheses, those hypotheses ($Ch(A) = 0.7$ and $Ch(A) = 0.4$) form a partition relative to E. Thus we can apply the Law of Total Probability (in its conditional credence form)[19] to obtain

$$Pr_H(A \mid E) = Pr_H(A \mid Ch(A) = 0.7 \ \& \ E) \cdot Pr_H(Ch(A) = 0.7 \mid E) + \\ Pr_H(A \mid Ch(A) = 0.4 \ \& \ E) \cdot Pr_H(Ch(A) = 0.4 \mid E) \qquad (5.5)$$

By Equations (5.3) and (5.4), this is

$$Pr_H(A \mid E) = 0.7 \cdot Pr_H(Ch(A) = 0.7 \mid E) + 0.4 \cdot Pr_H(Ch(A) = 0.4 \mid E) \quad (5.6)$$

As Equation (5.2) suggested, $Pr_H(\cdot \mid E)$ is just $cr(\cdot)$. So this last equation becomes

$$cr(A) = 0.7 \cdot cr(Ch(A) = 0.7) + 0.4 \cdot cr(Ch(A) = 0.4) \qquad (5.7)$$

Finally, we fill in the values stipulated in the problem to conclude

$$cr(A) = 0.7 \cdot 2/3 + 0.4 \cdot 1/3 = 0.6 \qquad (5.8)$$

That's a lot of calculation, but the overall lesson comes to this: When your total evidence is admissible and restricts you to a finite set of chance values for A, the Principal Principle sets your credence in A equal to a weighted average of those chance values (where each chance value is weighted by your credence that it's the true chance).

This is an extremely useful conclusion, *provided* we can discern what kinds of evidence are admissible. Lewis writes that, "Admissible propositions are the sort of information whose impact on credence about outcomes comes entirely by way of credence about the chances of those outcomes" (1980, p. 272). He then sketches out some categories of information that we should expect to be admissible, and inadmissible. For example, evidence about events causally upstream from the chances will be admissible; such events will form the first link in a causal chain like Figure 5.2. This includes information about the physical laws that give rise to chances—information that affects our credences about experimental outcomes by affecting our views about their chances. On the other hand, evidence about effects of the chance outcome is inadmissible, as we saw in the example of Amir's report. Generally, then, it's a good rule of thumb that facts concerning events temporally *before* the chance outcome are admissible, and inadmissible information is always about events *after* the outcome. (Though Lewis does remark at one point (1980, p. 274) that if backward causation is possible, seers of the future or time-travelers might give us inadmissible information about chance events yet to come.)

We'll close our discussion of the Principal Principle with a couple of caveats.[20] First, I have been talking about coin flips, die rolls, etc. as if their outcomes have non-extreme objective chances. If you think that these outcomes are fully determined by the physical state of the world prior to such events, you might think these examples aren't really chancy at all—or if there are chances associated with their outcomes, the world's determinism makes those chances either 1 or 0. There are authors who think non-extreme chance assignments are compatible with an event's being fully deterministic. This will be especially plausible if you think a single phenomenon may admit of causal explanations at multiple levels of description. (Though the behavior of a gas sample is fully determined by the positions and velocities of its constituent particles, we might still apply a statistical thermodynamics that treats the sample's behavior as chancy.) In any case, if the compatibility of determinism and non-extreme chance concerns you, you can replace my coin-flipping and die-rolling examples with genuinely indeterministic quantum events.

Second, you might think frequency data can affect rational credences without operating through opinions about chances. Suppose a new patient walks into a doctor's office, and the doctor assigns a credence that the patient has a particular disease equal to that disease's frequency in the general population.

In order for this to make sense, must the doctor assume that physical chances govern who gets the disease, or that the patient was somehow brought to her through a physically chancy process? (That is, must the frequency affect the doctor's credences by informing her opinions about chances?) This will depend on how broadly we are willing to interpret macroscopic events as having objective chances. But unless chances are literally everywhere, inferences governed by the Principal Principle form a proper subset of the legitimate instances of inductive reasoning. To move from frequencies in an observed population to predictions about the unobserved when chances are not present, we may need something like the frequency-credence principle (perhaps made more plausible by incorporating statistical tools) with which this section began. Or we may need a theory of inductive confirmation in general—something we will try to construct in Chapter 6.

For the time being, the message of the Principal Principle is clear: Where there are objective chances in the world, we should align our credences with them to the extent we can determine what they are. While there are exceptions to this rule, they can be worked out by thinking about the causal relations between our information and the chances of which we're aware.

5.2.2 Expert principles and Reflection

The Principal Principle is sometimes described as a **deference principle**: to the extent you can determine what the objective chances are, the principle directs you to defer to them by making your credences match. In a certain sense, you treat the chances as authorities on what your credences should be. Might other sorts of authorities demand such rational deference?

Testimonial evidence plays a large role in how we learn about the world. Suppose an expert on some subject reveals her credences to you. Instead of coming on television and talking about the "probability" of snow, the weather forecaster simply tells you she's 30% confident that it will snow tomorrow. It seems intuitive that—absent other evidence about tomorrow's weather—you should set your credence in snow to 0.30 as well.

We can generalize this intuition with a principle for deference to experts modeled on the Principal Principle:

$$\Pr_H(A \mid cr_E(A) = x) = x \tag{5.9}$$

Here Pr_H is a rational agent's hypothetical prior distribution, representing her ultimate epistemic standards for assigning attitudes on the basis of total evidence. A is a proposition within some particular subject matter, and $cr_E(A) = x$ is the proposition that an expert on that subject matter assigns credence x to A. As we've discussed before (Section 4.3), an agent's credences at a given time equal her hypothetical prior conditionalized on her total evidence at that time. So Equation (5.9) has consequences similar to the Principal Principle's: When a rational agent is *certain* that an expert assigns credence x to A, and that fact constitutes her total evidence relevant to A, satisfying Equation (5.9) will leave her with an unconditional credence of $cr(A) = x$. On the other hand, an agent who is uncertain of the expert's opinion can use Equation (5.9) to calculate a weighted average of all the values she thinks the expert might assign.[21]

Equation (5.9) helps us figure out how to defer to someone we've identified as an expert. But it doesn't say anything about how to make that identification! Ned Hall helpfully distinguishes two kinds of experts we might look for:

> Let us call the first kind of expert a *database-expert*: she earns her epistemic status simply because she possesses more information. Let us call the second kind an *analyst-expert*: she earns her epistemic status because she is particularly good at evaluating the relevance of one proposition to another.
>
> (2004, p. 100)

A **database expert**'s evidence (or at least, her evidence relevant to the matter at hand) is a superset of mine. While she may not reveal the contents of her evidence, I can still take advantage of it by assigning the credences she assigns on its basis. On the other hand, I defer to an **analyst expert** not because she has superior evidence but because she is particularly skilled at forming opinions from the evidence we share. Clearly these categories can overlap; relative to me, a weather forecaster is probably both an analyst expert and a database expert with respect to the weather.

One particular database expert has garnered a great deal of attention in the deference literature: an agent's future self. Because Conditionalization retains certainties (Section 4.1.1), at any given time a conditionalizing agent will possess all the evidence possessed by each of her past selves—and typically quite a bit more. So an agent who is certain she will update by conditionalizing should treat her future self as a database expert.[22] On the supposition that her future self will assign credence x to a proposition A, she should now assign credence x to A as well. This is van Fraassen's (1984)

Reflection Principle: For any proposition A in $\mathcal{L}$, real number x, and times t_i and t_j with $j > i$, rationality requires

$$cr_i(A \mid cr_j(A) = x) = x$$

Although the Reflection Principle mentions both the agent's t_i and t_j credences, strictly speaking it is a *synchronic* principle, relating various credences the agent assigns at t_i. If we apply the Ratio Formula and then cross-mutiply, Reflection gives us:

$$cr_i[A \ \& \ cr_j(A) = x] = x \cdot cr_i[cr_j(A) = x] \tag{5.10}$$

The two credences related by this equation are both assigned *at* t_i; they just happen to be credences *in* some propositions about t_j.

Despite this synchronic nature, Reflection bears an intimate connection to Conditionalization. If an agent is certain she will update by conditionalizing between t_i and t_j—and meets a few other side conditions—Reflection follows. For instance, the Reflection Principle can be proven from the following set of conditions:

1. The agent is certain at t_i that cr_j will result from conditionalizing cr_i on the total evidence she learns between t_i and t_j (call it E).
2. The agent is certain at t_i that E (whatever it may be) is true.
3. $cr_i(cr_j(A) = x) > 0$
4. At t_i the agent can identify a set of propositions S in $\mathcal{L}$ such that:
    (a) The members of S form a partition relative to the agent's certainties at t_i.
    (b) At t_i the agent is certain that E is one of the propositions in S.
    (c) For each member of S, the agent is certain at t_i what cr_j-value she assigns to A conditional on that member.

References to a proof can be found in the Further Reading. Here I'll simply provide an example that illustrates the connection between Conditionalization and Reflection. Suppose that I've rolled a die you're certain is fair, but as of t_1 have told you nothing about the outcome. However, at t_1 you're certain that between t_1 and t_2 I'll reveal to you whether the die came up odd or even. The Reflection Principle suggests you should assign

$$cr_1(6 \mid cr_2(6) = 1/3) = 1/3 \tag{5.11}$$

Assuming the enumerated conditions hold in this example, we can reason to Equation (5.11) as follows: In this case the partition S contains the proposition that the die came up odd and the proposition that it came up even. You are certain at t_1 that one of these propositions will provide the E you learn before t_2. You're also certain that your $cr_2(6)$ value will result from conditionalizing your t_1 credences on E. So you're certain at t_1 that

$$cr_2(6) = cr_1(6 \mid E) \qquad (5.12)$$

Equation (5.11) involves your t_1 credence in 6 conditional on the supposition that $cr_2(6) = 1/3$. To determine this value, let's see what conditional reasoning you could do at t_1, not yet certain what credences you will actually assign at t_2, but temporarily supposing that $cr_2(6) = 1/3$. We just said that at t_1 you're certain of Equation (5.12), so given the supposition you can conclude that $cr_1(6 \mid E) = 1/3$. Then you can examine your current t_1 credences conditional on both odd and even, and find that $cr_1(6 \mid E)$ will equal 1/3 only if E is the proposition that the die came up even. (Conditional on the die's coming up odd, your credence in a 6 would be 0.) Thus you can conclude that E is the proposition that the die came up even. You're also certain at t_1 that E (whatever its content) is true, so concluding that E says the die came up even allows you to conclude that the die did indeed come up even. And on the condition that the die came up even, your t_1 credence in a six is 1/3.

All of the reasoning in the previous paragraph was conditional, starting with the supposition that $cr_2(6) = 1/3$. We found that conditional on this supposition, your rational credence in six would be 1/3. And that's exactly what the Reflection Principle gave us in Equation (5.11).[23] Information about your future credences tells you something about what evidence you'll receive between now and then. And information about what evidence you'll receive in the future should be incorporated into your credences in the present.

But how often do we really get information about our future opinions? Approached the way I've just done, the Reflection Principle seems to have little real-world applicability. But van Fraassen originally proposed Reflection in a very different spirit. He saw the principle as stemming from basic commitments we undertake when we form opinions.

van Fraassen drew an analogy to making promises. Suppose I make a promise at a particular time, but at the same time admit to being unsure whether I will actually carry it out. van Fraassen writes that "To do so would mean that I am now less than fully committed (a) to giving due regard to the felicity conditions for this act, or (b) to standing by the commitments I shall overtly enter" (1984, p. 255). To fully stand behind a promise requires full confidence that you will carry it out. And what goes for current promises goes for future promises as well: if you know you'll make a promise later on, failing to be fully confident *now* that you'll enact the future promise is a betrayal of solidarity with your future promising self.

Now apply this lesson to the act of making judgments: assigning a different credence *now* to a proposition than the credence you know you'll assign in the future is a failure to stand by the commitments implicit in that future opinion. As van Fraassen puts it in a later publication, "Integrity requires me to express my commitment to proceed in what I now classify as a rational manner, to stand behind the ways in which I shall revise my values and opinions" (1995, pp. 25–6). This is his motivation for endorsing the Reflection Principle.[24] For van Fraassen, Reflection brings out a substantive commitment inherent in judgment, which underlies various other rational requirements. For instance, since van Fraassen's argument for Reflection does not *rely* on Conditionalization, van Fraassen at one point (1999) uses Reflection to *argue* for Conditionalization!

Of course, one might not agree with van Fraassen that assigning a credence involves such strong commitments. And even if Reflection can be supported as van Fraassen suggests, moving from that principle to Conditionalization is going to require substantive further premises. As we've seen, Reflection is a synchronic principle, relating an agent's attitudes at one time to other attitudes she assigns at the same time. By itself, Reflection may support a conclusion to the effect that an agent with certain attitudes at a given time is required to *predict* that she will update by Conditionalization. But to actually establish Conditionalization as a diachronic norm, we would need a further principle requiring rational agents to update in the manner they predict they will.[25]

5.3 The Principle of Indifference

The previous section discussed various deference principles (the Principal Principle, expert principles, the Reflection Principle) that place additional rational constraints on credence beyond the probability axioms, Ratio Formula, and Conditionalization. Yet each of those deference principles works with a particular kind of evidence—evidence about the chances, about an expert's credences, or about future attitudes. When an agent lacks these specific sorts of evidence about a proposition she's considering, the deference principles will do little to constrain her credences. If an Objective Bayesian (in the normative sense) wants to narrow what's rationally permissible to a single hypothetical prior, he is going to need a stronger principle than these.

The Principle of Indifference is often marketed to do the trick. This is John Maynard Keynes's name for what used to be known as the "principle of insufficient reason":

> The Principle of Indifference asserts that if there is no *known* reason for predicating of our subject one rather than another of several alternatives, then relatively to such knowledge the assertions of each of these alternatives have an *equal* probability.   (Keynes 1921, p. 42, emphasis in original)

Applied to degrees of belief, the **Principle of Indifference** holds that if an agent has no evidence favoring any proposition in a partition over any other, she should spread her credence equally over the members of the partition. If I tell you I have painted my house one of the seven colors of the rainbow but tell you nothing more about my selection, the Principle of Indifference requires you to assign credence 1/7 that my house is now violet.

The Principle of Indifference looks like it could settle all open questions about rational credence. An agent could assign specific credences as dictated by portions of her evidence (say, evidence that engages one of the deference principles), then use the Principle of Indifference to settle all remaining questions about her distribution. For example, suppose I tell you that I flipped a fair coin to decide on a house color—heads meant gray, while tails meant a color of the rainbow. You could follow the Principal Principle and assign credence 1/2 to my house's being gray, then follow the Principle of Indifference to distribute the remaining 1/2 credence equally among each of the rainbow colors (so each would receive credence 1/14). This plan seems to dictate a unique rational credence for every proposition in every evidential situation, thereby specifying a unique hypothetical prior distribution.

Unfortunately, the Principle of Indifference has a serious flaw, which was noted by Keynes (among others).[26] Suppose I tell you only that I painted my house some color—I don't tell you what palette I chose from—and you wonder whether it was violet. You might partition the possibilities into the proposition that I painted the house violet and the proposition that I didn't. In that case, the Principle of Indifference will require you to assign credence 1/2 that the house is violet. But if you use the seven colors of the rainbow as your partition, you will assign 1/7 credence that my house is now violet. And if you use the colors in a box of crayons.... The trouble is that faced with the same evidential situation and same proposition to be evaluated, the

Principle of Indifference will recommend different credences depending on which partition you consider.

Might one partition be superior to all the others, perhaps on grounds of the naturalness with which it divides the space of possibilities? (The selection of colors in a crayon box is pretty arbitrary!) Well, consider this example: I just drove eighty miles to visit you. I tell you it took between two and four hours to make the trip, and ask how confident you are that it took less than three. Three hours seems to neatly divide the possibilities in half, so by the Principle of Indifference you assign credence 1/2. Then I tell you I maintained a constant speed throughout the drive, and that speed was between 20 and 40 miles per hour. You consider the proposition that I drove faster than 30mph, and since that evenly divides the possible speeds the Indifference Principle again recommends a credence of 1/2. But these two credence assignments conflict. I drove over 30mph just in case it took me less than two hours and forty minutes to make the trip. So are you 1/2 confident that it took me less than three hours, or that it took me less than two hours forty minutes? If you assign any positive credence that my travel time fell between those durations, the two answers are inconsistent. So once more we need a specified partition (time or velocity) to apply the Principle of Indifference against. But here the decision can't be made on grounds of naturalness: thinking about one's speed of travel is neither more nor less natural than thinking about how long the trip took.[27]

This example is different from the painting example, in that time and velocity require us to consider continuous ranges of possibilities. Infinite possibility spaces introduce a number of complexities we will discuss in the next section, but hopefully the intuitive difficulty is clear. Joseph Bertrand (1888/1972) produced a number of infinite-possibility paradoxes for principles like Indifference. His most famous puzzle (now usually called **Bertrand's Paradox**) asks how probable it is that a chord of a circle will be longer than the side of an inscribed equilateral triangle. Indifference reasoning yields conflicting answers depending on how one specifies the chord in question—by specifying its endpoints, by specifying its orientation and length, by specifying its midpoint, etc.

Since Keynes's discussion, a number of authors have modified his Indifference Principle. Chapter 6 will look in detail at Carnap's proposal. Another well-known suggestion is E.T. Jaynes's (1957a,b) **Maximum Entropy Principle**. Given a partition of the space of possibilities, and a set of constraints on allowable credence distributions over that partition, the Maximum Entropy Principle selects the allowable distribution with the highest entropy. If the

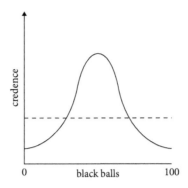

Figure 5.3 Possible urn distributions

partition is finite, containing the propositions $\{Q_1, Q_2, \ldots, Q_n\}$, the entropy of a distribution is calculated as

$$-\sum_{i=1}^{n} \mathrm{cr}(Q_i) \cdot \log \mathrm{cr}(Q_i) \tag{5.13}$$

The technical details of Jaynes's proposal are beyond the level of this book. The intuitive idea, though, is that by *maximizing* entropy in a distribution we *minimize* information.

To illustrate, suppose you know an urn contains 100 balls, each of which is either black or white. Initially, you assign an equal credence to each available hypothesis about how many black balls are in the urn. This "flat" distribution over the urn hypotheses is reflected by the dashed line in Figure 5.3. Then I tell you that the balls were created by a process that tends to produce roughly as many white balls as black. This moves you to the more "peaked" distribution of Figure 5.3's solid curve. The peaked distribution reflects the fact that at the later time you have more information about the contents of the urn. There are various mathematical ways to measure the informational content of a distribution, and it turns out that a distribution's entropy goes down as its information content goes up. So in Figure 5.3, the flat (dashed) distribution has a higher entropy than the peaked (solid) distribution.

Maximizing entropy is thus a strategy for selecting the lowest-information distribution consistent with what we already know. Jaynes's principle says that within the bounds imposed by your evidence, you should select the "flattest" credence distribution available. In a sense, this is a directive not to make any assumptions beyond what you know. As van Fraassen puts it, "one should not jump to unwarranted conclusions, or add capricious assumptions, when accommodating one's belief state to the deliverances of experience" (1981, p. 376). If *all* your evidence about my urn is that it contains 100 black or white balls, it would be strange for you to peak your credences around any particular number of black balls. What in your evidence would justify such a maneuver? The flat distribution seems the most rational option available.[28]

The Maximum Entropy approach has a number of advantages. First, it can easily be extended from finite partitions to infinite partitions by replacing the summation in Equation (5.13) with an integral (and making a few further adjustments). Second, for cases in which an agent's evidence simply delineates a space of doxastic possibilities (without favoring some of those possibilities over others), the Principle of Maximum Entropy yields the same results as the Principle of Indifference. But Maximum Entropy also handles cases involving more complicated sorts of information. Besides restricting the set of possibilities, an agent's evidence might require her credence in one possibility to be twice that in another, or might require a particular conditional credence value for some ordered pair of propositions. No matter the constraints, Maximum Entropy chooses the "flattest" (most entropic) distribution consistent with those constraints. Third, probability distributions selected by the Maximum Entropy Principle have been highly useful in various scientific applications, ranging from statistical mechanics to CT scans to natural language processing.

Yet the Maximum Entropy Principle also has flaws. It suffers from a version of the Indifference Principle's partitioning problem. Maximum Entropy requires us to first select a partition, then accept the most entropic distribution over that partition. But the probability value assigned to a particular proposition by this process often depends on what other propositions appear in the partition. Also, in some evidential situations satisfying the Maximum Entropy Principle both before and after an update requires agents to violate Conditionalization. You can learn more about these problems by studying this chapter's Further Reading.

5.4 Credences for infinitely many possibilities

Suppose I tell you a positive integer was just selected by some process, and tell you nothing more about that process. You need to distribute your credence across all the possible integers that might have been selected. Let's further suppose that you want to assign each positive integer the same credence. In the last section we asked whether, given your scant evidence in this case about the selection process, such an assignment is obligatory—whether you're rationally *required* to assign each positive integer an equal credence. In this section I want to set aside the question of whether an equal distribution is required, and ask whether it's even *possible*.

We're going to have a small, technical problem here with the propositional language over which your credence distribution is assigned. In Chapter 2 we set up propositional languages with a *finite* number of atomic propositions, while a distribution over every positive integer requires infinitely many atomic propositions. Yet there are standard logical methods for dealing with languages containing infinitely many atomic propositions, and even for representing them using a finite number of symbols. For example, we could use "1" to represent the atomic proposition that the number one was selected, "2" to represent two's being selected, "12" to represent twelve's being selected, etc. This will allow us to represent infinitely many atomic propositions with only the standard ten Arabic digits.

So the language isn't the real problem; the real problem is what single credence value you could possibly assign to each and every one of those positive integers. To start seeing the problem, imagine you pick some positive real number r and assign it as your unconditional credence in each positive integer's being picked. For any positive real r you pick, there exists an integer n such that $r > 1/n$. Select such an n, and consider the proposition that the positive integer selected was less than or equal to n. By Finite Additivity (Extended),

$$cr(1 \lor 2 \lor \ldots \lor n) = cr(1) + cr(2) + \ldots + cr(n) \qquad (5.14)$$

Each of the credences on the right-hand side equals r, so your credence in the disjunction is $r \cdot n$. But we selected n such that $r > 1/n$, so $r \cdot n > 1$. Which means the credence on the left-hand side of this equation is greater than 1, and you've violated the probability axioms.

This argument rules out assigning the same positive real credence to each and every positive integer. What other options are there? Historically the most popular proposal has been to assign each positive integer a credence of 0. Yet this proposal creates its own problems.

The first problem with assigning each integer zero credence is that we must reconceive what an unconditional credence of 0 means. So far in this book we have equated assigning credence 0 to a proposition with ruling that proposition out as a live possibility. In this case, though, we've proposed assigning credence 0 to each positive integer while still treating each as a live possibility. So while we will still assign credence 0 to propositions that have been ruled out, there will now be other types of propositions that receive credence 0 as well. Similarly, we may assign credence 1 to propositions of which we are not certain.

Among other things, this reconception of credence 0 will undermine arguments for the Regularity Principle. As stated (Section 4.2), Regularity forbids assigning credence 0 to any logically contingent proposition. The argument there was that one should never entirely rule out a proposition that's logically possible, so one should never assign such a proposition 0 credence. Now we've opened up the possibility of assigning credence 0 to a proposition without ruling it out. So while we can endorse the idea that no contingent proposition should be ruled out, Regularity no longer follows. Moreover, the current proposal provides infinitely many explicit counterexamples to Regularity: we have proposed assigning credence 0 to the contingent proposition that the positive integer selected was one, to the proposition that the integer was two, that it was three, etc.

Once we've decided to think about credence 0 in this new way, we encounter a second problem: the Ratio Formula. In Section 3.1.1 I framed the Ratio Formula as follows:

Ratio Formula: For any P and Q in $\mathcal{L}$, if $cr(Q) > 0$ then

$$cr(P \mid Q) = \frac{cr(P \,\&\, Q)}{cr(Q)}$$

This constraint relates an agent's conditional credence $cr(P \mid Q)$ to her unconditional credences *only when* $cr(Q) > 0$. As stated, it remains silent on how an agent's conditional and unconditional credences relate when $cr(Q) = 0$.

Yet we surely want to have some rational constraints on that relation for cases in which an agent assigns credence 0 to a contingent proposition that she hasn't ruled out.[29] For example, in the positive integer case consider your conditional credence $cr(2 \mid 2)$. Surely this conditional credence should equal 1. Yet because the current proposal sets $cr(2) = 0$, the Ratio Formula cannot tell us anything about $cr(2 \mid 2)$. And since we've derived all of our rational constraints on conditional credence from the Ratio Formula, the Bayesian system we've set up isn't going to deliver a requirement that $cr(2 \mid 2) = 1$.[30]

There are various ways to respond to this problem. One interesting suggestion is to reverse the order in which we proceeded with conditional and unconditional credences: We began by laying down fairly substantive constraints (Kolmogorov's probability axioms) on *un*conditional credences, then tied conditional credences to those via the Ratio Formula. On the reverse approach, substantive constraints are first placed on conditional credences, then some further rule relates unconditional to conditional. The simplest such rule is that for any proposition P, $cr(P) = cr(P \mid T)$.

Some advocates of this approach describe it as making conditional credence "basic", but we should be careful not to read too much into debates about what's basic. The way I've approached conditional and unconditional credences in this book, neither is more fundamental than the other in any sense significant to metaphysics or the philosophy of mind. Each is an independently existing type of doxastic attitude, and any rules we offer relating them are strictly *normative* constraints. The only sense in which our unconditionals-first approach has made unconditional credences prior to conditionals is in its order of normative explanation. The Ratio Formula helped us transform constraints on unconditional credences into constraints on conditional credences (as in Section 3.1.2). On the conditionals-first approach, the rule that $cr(P) = cr(P \mid T)$ transforms constraints on conditionals into constraints on unconditionals.

Examples of the conditionals-first approach include Hosiasson-Lindenbaum (1940), Popper (1955), Renyi (1970), and Roeper and Leblanc (1999).[31] Like many of these, Popper's axiom system entails that $cr(Q \mid Q) = 1$ for any Q that the agent deems possible, regardless of its unconditional credence value. This ensures that $cr(2 \mid 2) = 1$.

The final problem I want to address with assigning each positive integer 0 unconditional credence of being selected has to do with your unconditional credence that any integer was selected at all. The proposition that some integer was selected is equivalent to the disjunction of the proposition that one was selected, the proposition that two was selected, the proposition that three was selected, etc. Finite Additivity directly governs unconditional credences in disjunctions of two (mutually exclusive) disjuncts; iterating that rule gives us Finite Additivity (Extended), which applies to disjunctions of finitely many disjuncts. But this case concerns an *infinite* disjunction, and none of the constraints we've seen so far relates the unconditional credence of an infinite disjunction to the credences of its disjuncts.

It might seem natural to supplement our credence constraints with the following:

Countable Additivity: For any countable set $\{Q_1, Q_2, Q_3, \ldots\}$ of mutually exclusive propositions in $\mathcal{L}$,

$$cr(Q_1 \lor Q_2 \lor Q_3 \lor \ldots) = cr(Q_1) + cr(Q_2) + cr(Q_3) + \ldots$$

Notice that Countable Additivity does not apply to sets of *every* infinite size; it applies only to sets of *countably many* members. The set of positive integers is countable, while the set of real numbers is not. (If you are unfamiliar with the differing sizes of infinite sets, I would suggest studying the brief explanation referenced in this chapter's Further Reading.)

Countable Additivity naturally extends the idea behind Finite Additivity to sets of (countably) infinite size. Many authors have found it attractive. Yet in our example it rules out assigning credence 0 to each proposition stating that a particular positive integer was selected. Taken together, the proposition that one was selected, the proposition that two was selected, the proposition that three was selected, etc. form a countable set of mutually exclusive propositions (playing the role of Q_1, Q_2, Q_3, etc. in Countable Additivity). Countable Additivity therefore requires your credence in the disjunction of these propositions to equal the sum of your credences in the individual disjuncts. Yet the latter credences are each 0, while your credence in their disjunction (namely, the proposition that *some* positive integer was selected) should be 1.

So perhaps Countable Additivity wasn't such a good idea after all. The trouble is, without Countable Additivity we lose a very desirable property:

Conglomerability: For each proposition P and partition $\{Q_1, Q_2, Q_3, \dots\}$ in $\mathcal{L}$, $cr(P)$ is no greater than the largest $cr(P \mid Q_i)$ and no less than the least $cr(P \mid Q_i)$.

In other words, if Conglomerability holds then the largest $cr(P \mid Q_i)$ and the smallest $cr(P \mid Q_i)$ provide bounds between which $cr(P)$ must fall.

In defining Conglomerability I didn't say how large the Q-partitions in question are allowed to be. We might think of breaking up the general Conglomerability principle into a number of sub-cases: Finite Conglomerability applies to finite partitions, Countable Conglomerability applies to countable partitions, Continuous Conglomerability applies to partitions of continuum-many members, etc. Finite Conglomerability is guaranteed by the standard probability axioms. You'll prove this in Exercise 5.6, but the basic idea is that by the Law of Total Probability $cr(P)$ must be a weighted average of the various $cr(P \mid Q_i)$, so it can't be greater than the largest of them or less than the smallest. With the standard axioms in place, Countable Conglomerability then stands or falls with our decision about Countable Additivity; without Countable Additivity, Countable Conglomerability is false.[32]

We've already seen that the strategy of assigning 0 credence to each positive integer's being selected violates Countable Additivity; let's see how it violates

(Countable) Conglomerability as well.[33] Begin with the following definition: For any positive integer n that's not a multiple of 10, define the n-set as the set of all positive integers that start with n, followed by some number (perhaps 0) of zeroes. So the 1-set is $\{1, 10, 100, 1000, \ldots\}$; the 11-set is $\{11, 110, 1100, 11000, \ldots\}$; the 36-set is $\{36, 360, 3600, 36000, \ldots\}$; etc. Now take the proposition that the integer selected was a member of the 1-set, and the proposition that the integer selected was a member of the 2-set, and the proposition that the integer selected was a member of the 3-set, etc. (Though don't include any ns that are multiples of 10.) The set of these propositions forms a partition. (If you think about it carefully, you'll see that any positive integer that might have been selected belongs to exactly one of these sets.)

The distribution strategy we're considering is going to want to assign

$$cr(\text{the selected integer is not a multiple of 10} \mid$$
$$\text{the selected integer is a member of the 1-set}) = 0$$

$$(5.15)$$

Why is that? Well, the only number in the 1-set that is not a multiple of 10 is the number one. The 1-set contains infinitely many positive integers; on the supposition that one of those integers was selected you want to assign equal credence to each one's being selected; so you assign 0 credence to each one's being selected (including the number one) conditional on that supposition. This gives us Equation (5.15). The argument then generalizes; for any n-set you'll have

$$cr(\text{the selected integer is not a multiple of 10} \mid$$
$$\text{the selected integer is a member of that } n\text{-set}) = 0$$

$$(5.16)$$

Yet unconditionally it seems rational to have

$$cr(\text{the selected integer is not a multiple of 10}) = 9/10 \qquad (5.17)$$

Conditional on any particular member of our n-set partition, your credence that the selected integer isn't a multiple of 10 is 0. Yet unconditionally, you're highly confident that the integer selected is not a mutiple of 10. This is a flagrant violation of (Countable) Conglomerability—your credences in a particular proposition conditional on each member of a (countable) partition are all the same, yet your unconditional credence in that partition has a very different value!

Why is violating Conglomerability a problem? Well, imagine I'm about to give you some evidence on which you're going to conditionalize. In particular, I'm about to tell you to which of the n-sets the selected integer belongs. Whichever piece of evidence you're about to get, your credence that the integer isn't a multiple of 10 conditional on that evidence is 0. So you can be certain right now that immediately after receiving the evidence—whatever piece of evidence it turns out to be!—your credence that the integer isn't a multiple of 10 will be 0. Yet despite being certain that your better-informed future self will assign a particular proposition a credence of 0, you continue to assign that proposition a credence of 9/10 right now. This is a flagrant violation of the Reflection Principle, as well as generally good principles for attitude management. Our opinions are usually compromises among the pieces of evidence we think we might receive; we expect that some potential future pieces of evidence would change our views in one direction, while others would press in the other. If we know that no matter what evidence comes in we're going to be pulled away from our current opinion in the same direction, it seems irrationally stubborn to maintain our current opinion and not move in that direction right now. Conglomerability embodies these principles of good evidential hygiene; without Conglomerability our interactions with evidence begin to look absurd.

Where does this leave us? We wanted to find a way to assign an equal credence to each positive integer's being selected. We quickly concluded that that equal credence could not be a positive real number. So we considered assigning credence 0 to each integer's being selected. Doing so violates Countable Additivity (a natural extension of our finite principles for calculating credences in disjunctions) and Conglomerability, which looks desirable for a number of reasons. Are there any *other* options?

I will briefly mention two further possibilities. The first possibility is to assign each positive integer an **infinitesimal** credence of having been selected. To work with infinitesimals, we extend the standard real-number system to include numbers that are greater than 0 but smaller than all the positive reals. If we assign each integer an infinitesimal credence of having been picked, we avoid the problems with assigning a positive real and also the problems of assigning 0. (For instance, if you pile enough infinitesimals together they can sum to 1.) Yet infinitesimal numbers have a great deal of mathematical structure, and it's not clear that the extra structure plausibly represents any feature of agents' attitudes.[34] Moreover, the baroque mathematics of infinitesimals introduces troubles of its own (see Further Reading). So perhaps only one viable option remains: Perhaps if you learn a positive integer was just selected, it's *impossible* to assign equal credence to each of the possibilities consistent with what you know.[35]

5.5 Jeffrey Conditionalization

Section 4.1.1 showed that conditionalizing on new evidence creates and retains certainties; evidence gained between two times becomes certain at the later time and remains so ever after. Contraposing, if an agent updates by Conditionalization and gains no certainties between two times, it must be because she gained no evidence between those times. In that section we also saw that if an agent gains no evidence between two times, Conditionalization keeps her credences fixed. Putting all this together, we see that under Conditionalization an agent's credences change just in case she gains new certainties.

As we noted in Section 4.2, mid-twentieth-century epistemologists like C.I. Lewis defended this approach by citing sense data as the foundational evidential certainties. Many contemporary epistemologists are uncomfortable with this kind of foundationalism (and with appeals to sense data in general). Richard C. Jeffrey, however, had a slightly different concern, which he expressed with the following example and analysis:

> The agent inspects a piece of cloth by candlelight, and gets the impression that it is green, although he concedes that it might be blue or even (but very improbably) violet. If G, B, and V are the propositions that the cloth is green, blue, and violet, respectively, then the outcome of the observation might be that, whereas originally his degrees of belief in G, B, and V were .30, .30, and .40, his degrees of belief in those same propositions after the observation are .70, .25, and .05. If there were a proposition E in his preference ranking which described the precise quality of his visual experience in looking at the cloth, one would say that what the agent learned from the observation was that E is true....

> But there need be no such proposition E in his preference ranking; nor need any such proposition be expressible in the English language. Thus, the description "The cloth looked green or possibly blue or conceivably violet," would be too vague to convey the precise quality of the experience. Certainly, it would be too vague to support such precise conditional probability ascriptions as those noted above. It seems that the best we can do is to describe, not the quality of the visual experience itself, but rather its effects on the observer, by saying, "After the observation, the agent's degrees of belief in G, B, and V were .70, .25, and .05."   (1965, p. 154)

Jeffrey worried that even if we grant the existence of a sense datum for each potential learning experience, the quality of that sense datum might not be representable in a proposition to which the agent could assign certainty, or at least might not be representable in a precise-enough proposition to differentiate that sense datum from other nearby data with different effects on the agent's credences.

At the time Jeffrey was writing, the standard Bayesian updating norm (updating by Conditionalization) relied on the availability of such propositions. So Jeffrey proposed a new updating rule, capable of handling examples like the cloth one above. While he called it **probability kinematics**, it is now universally known as

Jeffrey Conditionalization: Given any t_i and t_j with $i < j$, any A in $\mathcal{L}$, and a finite partition $\{B_1, B_2, \ldots, B_n\}$ in $\mathcal{L}$ whose members each have nonzero cr_i,

$$cr_j(A) = cr_i(A \mid B_1) \cdot cr_j(B_1) + cr_i(A \mid B_2) \cdot cr_j(B_2) + \ldots + cr_i(A \mid B_n) \cdot cr_j(B_n)$$

Let's apply Jeffrey Conditionalization to the cloth example. Suppose I'm fishing around in a stack of my family's clean laundry hoping to pull out any shirt that belongs to me, but the lighting is dim because I don't want to turn on the overheads and awaken my wife. The color of a shirt in the stack would be a strong clue as to whether it was mine, as reflected by my conditional credences:

$$cr_1(\text{mine} \mid G) = 0.80$$
$$cr_1(\text{mine} \mid B) = 0.50 \qquad (5.18)$$
$$cr_1(\text{mine} \mid V) = 0$$

(For simplicity's sake we imagine green, blue, and violet are the only color shirts I might fish out of the stack.) At t_1 I pull out a shirt. Between t_1 and t_2 I take a glimpse of the shirt. According to Jeffrey's story, my unconditional credence distributions across the $G/B/V$ partition are:

$$cr_1(G) = 0.30 \qquad cr_1(B) = 0.30 \qquad cr_1(V) = 0.40$$
$$cr_2(G) = 0.70 \qquad cr_2(B) = 0.25 \qquad cr_2(V) = 0.05 \qquad (5.19)$$

Applying Jeffrey Conditionalization, I find my credence in the target proposition at the later time by combining my post-update unconditional credences across the partition with my pre-update credences in the target proposition conditional on members of the partition. This yields:

$$cr_2(\text{mine}) =$$
$$cr_1(\text{mine} \mid G) \cdot cr_2(G) + cr_1(\text{mine} \mid B) \cdot cr_2(B) + cr_1(\text{mine} \mid V) \cdot cr_2(V) =$$
$$0.80 \cdot 0.70 + 0.50 \cdot 0.25 + 0 \cdot 0.05 =$$
$$0.685$$

$$(5.20)$$

At t_2 I'm fairly confident that the shirt I've selected is mine. How confident was I at t_1, before I caught my low-light glimpse? A quick calculation with the Law of Total Probability reveals that $cr_1(\text{mine}) = 0.39$. But it's more interesting to see what happens when we apply the Law of Total Probability to my credences at t_2:

$$cr_2(\text{mine}) =$$
$$cr_2(\text{mine} \mid G) \cdot cr_2(G) + cr_2(\text{mine} \mid B) \cdot cr_2(B) + cr_2(\text{mine} \mid V) \cdot cr_2(V)$$
$$(5.21)$$

Take a moment to compare Equation (5.21) with the first two lines of Equation (5.20). Equation (5.21) expresses a feature that my t_2 credence distribution must have if it is to satisfy the probability axioms and Ratio Formula. Equation (5.20) tells me how to set my t_2 credences by Jeffrey Conditionalization. The only way to make these two equations match—the only way to square the Jeffrey update with the probability calculus—is if $cr_1(\text{mine} \mid G) = cr_2(\text{mine} \mid G)$, $cr_1(\text{mine} \mid B) = cr_2(\text{mine} \mid B)$, etc.

Why should these conditional credences stay constant over time? Well, at any given time my credence that the shirt I've selected is mine is a function of two kinds of credences: first, my unconditional credence that the shirt is a particular color; and second, my conditional credence that the shirt is mine given that it's a particular color. When I catch a glimpse of the shirt between t_1 and t_2, *only the first kind of credence changes*. I change my opinion about what color the shirt is, but I don't change my confidence that it's my shirt given that (say) it's green. Throughout the example I have a fixed opinion about what percentage of the green shirts in the house are mine; I simply gain information about whether *this* shirt is green. So while my unconditional color credences change, my credences conditional on the colors remain.

This discussion reveals a general feature of Jeffrey Conditionalization. You'll prove in Exercise 5.8 that an agent's credences between two times update by Jeffrey Conditionalzation just in case the following condition obtains:

Rigidity: For any A in $\mathcal{L}$ and any B_m in $\{B_1, B_2, \ldots, B_n\}$,

$$cr_j(A \mid B_m) = cr_i(A \mid B_m)$$

So Jeffrey Conditionalization using a particular partition $\{B_1, B_2, \ldots, B_n\}$ is appropriate only when the agent's credences conditional on the B_m remain constant across two times. Jeffrey thought this was reasonable for updates that "originate" in the B_m partition.[36] In the cloth example, all my credence changes between t_1 and t_2 are *driven* by the changes in my color credences caused by my experience. So if I tell you my credences at t_1, and then tell you my unconditional credences in the color propositions at t_2, this should suffice for you to work out the rest of my opinions at t_2. Jeffrey Conditionalization makes that possible.

Rigidity can help us perform Jeffrey Conditionalization updates on a proba-bility table. Given the partition $\{B_1, B_2, \ldots, B_n\}$ in which an update originates, we divide the lines of the table into "blocks": the B_1 block contains all the lines consistent with B_1; the B_2 block contains all the lines consistent with B_2; etc. The agent's experience between times t_i and t_j directly sets her unconditional cr_j-values for the B_m; in other words, it tells us what each block must sum to at t_j. Once we know a block's cr_j total, we set the values on individual lines within that block by keeping them in the same proportions as at t_i. (This follows from Rigidity's requirement that each line have the same cr_2-value conditional on a given B_m as it did at t_i.) That is, we multiply all the cr_i-values in a block by the *same* constant so that their cr_j-values achieve the appropriate sum.

Figure 5.4 shows this process for the colored shirt example. I've built the table around a simplified partition of doxastic possibilities in the problem, but I could've made a probability table with the full list of state-descriptions and everything would proceed the same way. I calculated the cr_1-values in the table from Equations (5.18) and (5.19). How do we then derive the credences at t_2?

The credence change between t_1 and t_2 originates in the $G/B/V$ partition. So the "blocks" on this table will be pairs of adjacent lines: the first pair of lines (on which G is true), the second pair of lines (B lines), and the third pair of

partition member	cr_1	cr_2
G & mine	0.24	0.56
G & ~mine	0.06	0.14
B & mine	0.15	0.125
B & ~mine	0.15	0.125
V & mine	0	0
V & ~mine	0.40	0.05

Figure 5.4 Jeffrey Conditionalization across a partition

V lines. Let's work with the B-block first. In Jeffrey's story, glimpsing the shirt sends me to $cr_2(B) = 0.25$. So on the table, the third and fourth lines must have cr_2-values summing to 0.25. At t_1 these lines were in a $1:1$ ratio, so they must maintain that ratio at t_2. This leads to cr_2-values of 0.125 on both lines. Applying a similar process to the G- and V-blocks yields the remaining cr_2-values.

Once you understand this block-updating process, you can see that traditional updating by Conditionalization is a special case of updating by Jeffrey Conditionalization. When you update by Conditionalization on some evidential proposition E, your probability table divides into two blocks: lines consistent with E versus $\sim E$ lines. After the update, the $\sim E$ lines go to zero, while the E lines are multiplied by a constant so that they sum to 1.

This tells us how Jeffrey Conditionalization relates to traditional (or "strict") Conditionalization mathematically. But how should we understand their relation philosophically? Suppose we class learning experiences into two kinds: those that send some proposition to certainty and those that don't. Jeffrey Conditionalization seems to be a universal updating rule, applying to both kinds of experience. When experience does send a proposition to certainty, Jeffrey Conditionalization provides the same advice as strict Conditionalization. But Jeffrey Conditionalization also provides guidance for learning experiences of the second kind.

Now the defender of Regularity (the principle forbidding extreme unconditional credence in logically contingent propositions) will maintain that *only* the second kind of learning experience ever occurs (at least to rational agents), and therefore that strict Conditionalization should *never* be applied in practice. All experience ever does is shuffle an agent's unconditional credences over some partition, without sending any partition members to extremity. Jeffrey Conditionalization tells us how such changes over a partition affect the rest of the agent's credence distribution.

But one can identify an important role for Jeffrey Conditionalization even without endorsing Regularity. To establish the need for his new kinematics, Jeffrey only had to argue that *some* experiences of the second kind exist— *sometimes* we learn without gaining certainties. In that case we need a more general updating rule than strict Conditionalization, and Jeffrey Conditionalization provides one.

Yet despite being such a flexible tool, Jeffrey Conditionalization has its drawbacks. For instance, while applications of strict Conditionalization are always commutative, Jeffrey updates that do not send propositions to certainty may not be. The simplest example of this phenomenon (which Jeffrey readily acknowledged) occurs when one learning experience sends some B_m in the partition to unconditional credence p, while the next experience sends that same partition member to a different credence value q. Applying Jeffrey Conditionalization to the experiences in that order will leave the agent with a final unconditional credence in B_m of q, while applying Jeffrey's rule to the same experiences in the opposite order will result in a final B_m credence of p. This commutativity failure is problematic if you think that the effects of evidence on an agent should not depend on the order in which pieces of evidence arrive.[37]

Finally, Jeffrey Conditionalization may not provide a recipe for *every* type of learning experience. Traditional Conditionalization covers experiences that set unconditional credences to certainty. Jeffrey Conditionalization generalizes to experiences that set unconditional credences to nonextreme values. But what if an experience affects an agent by directly altering her *conditional* credences? How can we calculate the effects of such an experience on her other degrees of belief? Readers interested in that question might begin by exploring van Fraassen's "Judy Benjamin Problem" (1981), an example in which direct alteration of conditional credences plausibly occurs, but which cannot be analyzed using Jeffrey Conditionalization.[38]

5.6 Exercises

Unless otherwise noted, you should assume when completing these exercises that the credence distributions under discussion satisfy the probability axioms and Ratio Formula. You may also assume that whenever a conditional credence expression occurs, the needed proposition has nonzero unconditional credence so that conditional credences are well defined.

Problem 5.1. 🎵 At noon I rolled a six-sided die. It came from either the Fair Factory (which produces exclusively fair dice), the Snake-Eyes Factory (which produces dice with a 1/2 chance of coming up one and equal chance of each other outcome), or the Boxcar Factory (which produces dice with a 1/4 chance of coming up six and equal chance of each other outcome).

    (a) Suppose you use the Principle of Indifference to assign equal credence to each of the three factories from which the die might have come. Applying the Principal Principle, what is your credence that my die roll came up three?

    (b) Maria tells you that the die I rolled didn't come from the Boxcar Factory. If you update on this new evidence by Conditionalization, how confident are you that the roll came up three?

    (c) Is Maria's evidence admissible with respect to the outcome of the die roll? Explain.

    (d) After you've incorporated Maria's information into your credence distribution, Ron tells you the roll didn't come up six. How confident are you in a three after conditionalizing on Ron's information?

    (e) Is Ron's evidence admissible with respect to the outcome of the die roll? Explain.

Problem 5.2. ✏ The expert deference principle in Equation (5.9) resembles the Principal Principle in many ways. Yet the expert deference principle makes no allowance for anything like inadmissible information. What kind of information should play the role for expert deference that inadmissible information plays for deference to chances? How should Equation (5.9) be modified to take such information into account?

Problem 5.3. 🎵🎵 Suppose t_1, t_2, and t_3 are three times, with the indices reflecting their temporal order. At t_1, you satisfy the probability axioms, Ratio Formula, and Reflection Principle. You are also certain at t_1 that you will satisfy these constraints at t_2. However, for some proposition X your t_1 credences are equally divided between the following two (mutually exclusive and exhaustive) hypotheses about what your t_2 self will think of your t_3 credences:

    Y: $(cr_2[cr_3(X) = 1/10] = 1/3) \& (cr_2[cr_3(X) = 2/5] = 2/3)$
    Z: $(cr_2[cr_3(X) = 3/8] = 3/4) \& (cr_2[cr_3(X) = 7/8] = 1/4)$

Given all this information, what is $cr_1(X)$? (Be sure to explain your reasoning clearly.)

Problem 5.4. ✒ Can you think of any kind of real-world situation in which it would be rationally permissible to violate the Reflection Principle? Explain the situation you're thinking of, and why it would make a Reflection violation okay.

Problem 5.5. ✎ Jingyi assigns the t_1 credences indicated by the probability table below. Then between t_1 and t_2, she learns $P \supset Q$.

P	Q	cr_1
T	T	0.4
T	F	0.2
F	T	0.2
F	F	0.2

(a) Determine Jingyi's credence distribution at t_2. Then use Equation (5.13) to calculate the entropy of both cr_1 and cr_2 over the partition containing the four P/Q state-descriptions.[39]
(b) Use the concept of information content to explain why the entropy of Jingyi's distribution changed in the direction it did between t_1 and t_2.
(c) Create a probabilistic credence distribution that assigns the same unconditional value to P as cr_1, but has a higher entropy over the P/Q state-description partition.
(d) Use the partition containing just P and $\sim P$ to calculate the entropy for cr_1 and for your distribution from part (c). What does this tell you about the partition-dependence of entropy comparisons?

Problem 5.6. ✎✎ Using Non-Negativity, Normality, Finite Additivity, the Ratio Formula, and any results we've proven from those four, prove Finite Conglomerability. (<u>Hint</u>: The Law of Total Probability may be useful here.)

Problem 5.7. ✎ Suppose that at t_1 you assign a "flat" credence distribution over language $\mathcal{L}$ whose only two atomic propositions are B and C—that is, you assign equal credence to each of the four state-descriptions of $\mathcal{L}$. Between t_1 and t_2 you perform a Jeffrey Conditionalization that originates in the $B/\sim B$ partition and sets $cr_2(B) = 2/3$. Between t_2 and t_3 you perform a Jeffrey Conditionalization that originates in the $C/\sim C$ partition and sets $cr_3(C) = 3/4$.
(a) Calculate your cr_2 and cr_3 distributions.
(b) Does your credence in B change between t_2 and t_3? Does your credence in C change between t_1 and t_2?

(c) By talking about probabilistic independence at t_1 and t_2, explain the changes or lack of changes you observed in parts (b) and (c).

(d) Now start again with the flat t_1 distribution, but apply the Jeffrey Conditionalizations in the opposite order. (First an update that sets the C credence to 3/4, then an update that sets B to 2/3.)

(e) Is the cr_3 distribution you obtained in part (e) the same as the one from part (a)? Does this always happen when you reverse the order of Jeffrey Conditionalizations? If not, why do you think it happened in this case?

Problem 5.8. 𝄞𝄞 Prove that Jeffrey Conditionalization is equivalent to Rigidity. That is: Given any times t_i and t_j, and any finite partition $\{B_1, B_2, \ldots, B_n\}$ in $\mathcal{L}$ whose members each have nonzero cr_i, the following two conditions are equivalent:

1. For all A in $\mathcal{L}$, $cr_j(A) = cr_i(A \mid B_1) \cdot cr_j(B_1) + cr_i(A \mid B_2) \cdot cr_j(B_2) + \ldots + cr_i(A \mid B_n) \cdot cr_j(B_n)$.
2. For all A in $\mathcal{L}$ and all B_m in the partition, $cr_j(A \mid B_m) = cr_i(A \mid B_m)$.

(Hint: Complete two proofs—first condition 2 from condition 1, then vice versa.)

Problem 5.9. 𝄞𝄞𝄞 Suppose we apply Jeffrey Conditionalization over a finite partition $\{B_1, B_2, \ldots, B_n\}$ in $\mathcal{L}$ to generate cr_2 from cr_1. Show that we could have obtained the same cr_2 from cr_1 in the following way: start with cr_1; Jeffrey Conditionalize it in a particular way over a partition containing only two propositions; Jeffrey Conditionalize the result of *that* operation in a particular way over a partition containing only two propositions (possibly different from the ones used the first time); repeat this process a finite number of times until cr_2 is eventually obtained.[40]

5.7 Further reading

SUBJECTIVE AND OBJECTIVE BAYESIANISM

Maria Carla Galavotti (2005). *Philosophical Introduction to Probability*. CSLI Lecture Notes 167. Stanford, CA: CSLI Publications

Excellent historical introduction to the many ways "probability" has been understood by the philosophical and statistical community.

Alan Hájek (2019). Interpretations of Probability. In: *The Stanford Encyclopedia of Philosophy*. Ed. by Edward N. Zalta. Fall 2019. URL: http://plato.stanford.edu/archives/fall2019/entries/probability-interpret/

Survey of the various interpretations of probability, with extensive references.

Bruno de Finetti (1931/1989). Probabilism: A Critical Essay on the Theory of Probability and the Value of Science. *Erkenntnis* 31, pp. 169–223. (Translation of B. de Finetti, *Probabilismo*, Logos 14: 163–219)

Classic paper critiquing objective interpretations of probability and advocating a Subjective Bayesian (in the semantic sense) approach.

Donald Gillies (2000). Varieties of Propensity. *British Journal for the Philosophy of Science* 51, pp. 807–35

Reviews different versions of the propensity theory and their motivations. Focuses at the end on how propensity theories might respond to Humphreys's Paradox.

Deference Principles

David Lewis (1980). A Subjectivist's Guide to Objective Chance. In: *Studies in Inductive Logic and Probability*. Ed. by Richard C. Jeffrey. Vol. 2. Berkeley: University of California Press, pp. 263–94

Lewis's classic article laying out the Principal Principle and its consequences for theories of credence and chance.

Adam Elga (2007). Reflection and Disagreement. *Noûs* 41, pp. 478–502

Offers principles for deferring to many different kinds of agents, including experts, gurus (individuals with good judgment who lack some of your evidence), past and future selves, and peers (whose judgment is roughly as good as your own).

Bas C. van Fraassen (1984). Belief and the Will. *The Journal of Philosophy* 81, pp. 235–56

Article in which van Fraassen proposes and defends the Reflection Principle.

Jonathan Weisberg (2007). Conditionalization, Reflection, and Self-Knowledge. *Philosophical Studies* 135, pp. 179–97

Discusses conditions under which Reflection can be derived from Conditionalization, and vice versa.

Richard Pettigrew and Michael G. Titelbaum (2014). Deference Done Right. *Philosophers' Imprint* 14, pp. 1–19

Attempts to get the formulation of deference principles precisely right, including expert deference principles, the Reflection Principle, and principles for higher-order credences. Particularly concerned with making those principles consistent with Conditionalization and with the possibility of ignorance about what's rationally required.

The Principle of Indifference

John Maynard Keynes (1921). *Treatise on Probability*. London: Macmillan and Co., Limited

Chapter IV contains Keynes's famous discussion of the Principle of Indifference.

E. T. Jaynes (1957a). Information Theory and Statistical Mechanics I. *Physical Review* 106, pp. 620–30
E. T. Jaynes (1957b). Information Theory and Statistical Mechanics II. *Physical Review* 108, pp. 171–90

E.T. Jaynes introduces the Maximum Entropy approach.

Colin Howson and Peter Urbach (2006). *Scientific Reasoning: The Bayesian Approach*. 3rd edition. Chicago: Open Court

Section 9.a covers the Indifference Principle, Harold Jeffreys's attempts to make it partition-invariant, and then Jaynes's Maximum Entropy theory. Very clear on the flaws of all of these approaches.

Teddy Seidenfeld (1986). Entropy and Uncertainty. *Philosophy of Science* 53, pp. 467–91

A general discussion of the flaws with Jaynes's Maximum Entropy approach; especially good on its incompatibility with Bayesian conditionalization. Also contains useful references to Jaynes's many defenses of Maximum Entropy over the years and to the critical discussion that has ensued.

CREDENCES FOR INFINITE POSSIBILITIES

David Papineau (2012). *Philosophical Devices: Proofs, Probabilities, Possibilities, and Sets.* Oxford: Oxford University Press

Chapter 2 offers a highly accessible introduction to the cardinalities of various infinite sets. (Note that Papineau uses "denumerable" where we use the term "countable".)

Alan Hájek (2003). What Conditional Probability Could Not Be. *Synthese* 137, pp. 273–323

Assesses the viability of the Ratio Formula as a definition of conditional probability in light of various infinite phenomena and plausible violations of Regularity.

Colin Howson (2014). Finite Additivity, Another Lottery Paradox and Conditionalisation. *Synthese* 191, pp. 989–1012

Neatly surveys arguments for and against Countable Additivity, then argues for dropping Conditionalization as a universal update rule over accepting infinite additivity principles.

Timothy Williamson (2007). How Probable Is an Infinite Sequence of Heads? *Analysis* 67, pp. 173–80

Brief introduction to the use of infinitesimals in probability distributions, followed by an argument against using infinitesimals to deal with infinite cases.

Kenny Easwaran (2014b). Regularity and Hyperreal Credences. *Philosophical Review* 123, pp. 1–41

Excellent, comprehensive discussion of the motivations for Regularity, the mathematics of infinitesimals, arguments against using infinitesimals to secure Regularity (including Williamson's argument), and an alternative approach.

Jeffrey Conditionalization

> Richard C. Jeffrey (1965). *The Logic of Decision*. 1st edition. McGraw-Hill Series in Probability and Statistics. New York: McGraw-Hill

Chapter 11 contains Jeffrey's classic presentation of his "probability kinematics", now universally known as "Jeffrey Conditionalization".

Notes

1. The frequency theory is sometimes referred to as "frequentism" and its adherents as "frequentists". However "frequentism" more often refers to a school of statistical practice at odds with Bayesianism (which we'll discuss in Chapter 13). The ambiguity probably comes from the fact that most people in that statistical school also adopt the frequency theory as their interpretation of probability. But the positions are logically distinct and should be denoted by different terms. So I will use "frequency theory" here, and reserve "frequentism" for my later discussion of the statistical approach.
2. For many, many more see Hájek (1996) and its sequel Hájek (2009b).
3. The frequency theory will also need to work with counterfactuals if nonextreme probabilities can be meaningfully ascribed to a priori truths, or to metaphysical necessities. (Might a chemist at some point have said, "It's highly probable that water is H_2O"?) Assigning nonextreme frequencies to such propositions' truth involves possible worlds far away from the actual.
4. This difficulty for the propensity theory is often known as **Humphreys's Paradox**, since it was proposed in Humphreys (1985).

   One might respond to Humphreys's Paradox by suggesting that propensities don't follow the standard mathematical rules of probability. And honestly, it's not obvious why they should. The frequency theory clearly yields probabilistic values: in any sequence of event repetitions a given outcome has a non-negative frequency, the tautologous outcome has a frequency of 1, and mutually exclusive outcomes have frequencies summing to the frequency of their disjunction. In fact, Kolmogorov's axioms can be read as a generalization of the mathematics of event frequencies to cases involving irrational and infinite quantities. But establishing that propensity values (or objective chances) satisfy the probability axioms takes *argumentation* from one's metaphysics of propensity. Nevertheless, most authors who work with propensities assume that they satisfy the axioms; if they didn't, the propensity interpretation's probabilities wouldn't count as probabilities in the mathematician's sense (Section 2.2).

5. One could focus here on a metaphysical distinction rather than a semantic one—instead of asking what "probability" talk *means*, I could ask what probabilities *are*. But some of the probability interpretations we will discuss don't have clear metaphysical commitments. The logical interpretation, for instance, takes probability to be a logical relation, but need not go on to specify an ontology for such relations. So I will stick with a semantic distinction, which in any case matches how these questions were discussed in much of twentieth-century analytic philosophy.

6. In the twentieth century Subjective Bayesianism was also typically read as a form of expressivism; an agent's "probability" talk *expressed* her quantitative attitudes toward propositions without having truth-conditions. Nowadays alternative semantics are available that could interpret "probability" talk in a more cognitivist mode while still reading such talk as reflecting subjective degrees of belief (Weatherson and Egan 2011).

7. See Hacking (1971) for discussion of Leibniz's position.

8. Carnap himself did not believe all "probability" talk picked out the logical values just described. Instead, he thought "probability" was ambiguous between two meanings, one of which was logical probability and the other of which had more of a frequency interpretation.

9. There is disagreement about whether the logical and evidential interpretations of probability should be considered Objective Bayesian in the semantic sense. Popper (1957) says that objective interpretations make probability values objectively *testable*. Logical and evidential probabilities don't satisfy that criterion, and Popper seems to class them as subjective interpretations. Yet other authors (such as Galavotti 2005) distinguish between logical and subjective interpretations. I have defined the semantic Subjective/Objective Bayesian distinction so that logical and evidential interpretations count as Objective; while they may be normative for the attitudes of agents, logical and evidential probabilities do not vary with the attitudes particular agents or groups of agents possess.

10. As I explained in Chapter 4, note 17, defining hypothetical priors as regular does not commit us to the Regularity Principle as a rational constraint.

11. Those who believe that "probability" is used in many ways—or that there are many different kinds of entities that count as probabilities—sometimes use the terms "subjective probability" and "objective probability". On this usage, subjective probabilities are agents' credences, while objective probabilities include all the kinds of probabilities we've mentioned that are independent of particular agents' attitudes.

12. To assign H a credence exactly equal to the observed frequency of heads would be to follow what Reichenbach (1938) called the **straight rule**. Interestingly, it's impossible to construct a hypothetical prior satisfying the probability axioms that allows an agent to obey the straight rule in its full generality. However, Laplace (1814/1995) proved that if an agent's prior satisfies the Principle of Indifference (adopting a "flat" distribution somewhat like the dashed line in Figure 5.3), her posteriors will obey the **rule of succession**: after seeing h of n tosses come up heads, her credence in H will be $(h+1)/(n+2)$. As the number of tosses increases, this credence approaches the observed frequency of heads.

Given these difficulties aligning credences and observed frequencies, anyone who thinks credences should match chances needs to describe a hypothetical prior making such a match possible. In a moment we'll see Lewis doing this with the Principal Principle.

13. Since the ratio of *B*-outcomes to *A*-events must always fall between 0 and 1, this principle sheds some light on why credence values are usually scaled from 0 to 1. (Compare note 4 above.)

14. There's also the problem that we sometimes have data from overlapping reference classes applying to the same case, neither of which is a subclass of the other. *The Book of Odds* (Shapiro, Campbell, and Wright 2014, p. 137) reports that 1 in 41.7 adults in the United States aged 20 or older experiences heart failure in a given year. For non-Hispanic white men 20 or older, the number is 1 in 37. But only 1 in 500 men aged 20–39 experiences heart failure in a given year. In setting my credence that I will have a heart attack this year, should I use the data for non-Hispanic white men over 20 or the data for men aged 20–39?

15. Here I'm thinking especially of the following: "What makes it be so that a certain reasonable initial credence function and a certain reasonable system of basic intrinsic values are both yours is that you are disposed to act in more or less the ways that are rationalized by the pair of them together, taking into account the modification of credence by conditionalizing on total evidence" (Lewis 1980, p. 288).

16. My explanation at this point in the text of screening-off in the Principal Principle fits very naturally with a propensity-style account of chance. I'm unsure whether it could be made to work on Lewis's own "best system" theory of chance (Lewis 1994). As far as I know, Lewis himself never explains why the screening-off captured by the Principal Principle should obtain, except to say that it matches our best intuitions about how rational agents assign credences to chance events.

17. The notion of screening off in play here is the one I described in Chapter 3, note 9 for continuous random variables. The objective chance of *H* is a continuous variable, so facts about $\mathrm{Ch}(H)$ screen off known flip frequencies from *H* in the sense that conditional on setting $\mathrm{Ch}(H)$ to any particular value, known frequency information becomes irrelevant to *H*.

18. Notice that the time t_i to which the chance in the Principal Principle is indexed need not be the time at which an agent assigns her credence concerning the experimental outcome *A*. In our coin example, the agent forms her credence at 1 p.m. about the coin flip outcome at noon using information about the chances *at noon*. This is significant because on some metaphysical theories of chance, once the coin flip lands heads (or tails) the chance of *H* goes to 1 (or 0) forevermore. Yet even if the chance of *H* has become extreme by 1 p.m., the Principal Principle may still direct an agent to assign a nonextreme 1 p.m. credence to *H* if all she knows are the chances from an earlier time. (Getting this last point wrong is the most frequent mistake I see people make in applying the Principal Principle. For more such mistakes, see Meacham 2010b.)

I should also note that because chances are time-indexed, the notion of admissibility must be time-indexed as well. The information about the wad of chewing gum is admissible relative to 11:30 a.m. chances—learning about the chewing gum affects your credence about the flip outcome by way of your opinions about the 11:30 a.m. chances. But the information that chewing gum was stuck to the coin after 11 a.m. is *in*admissible relative to the 11 a.m. chances. (Chewing gum information affects your credence in *H*, but not by influencing your opinions about the chances associated with the coin at 11 a.m.) So strictly speaking we should ask whether a piece of information is admissible *for* a particular proposition *relative* to the chances at a given time. I have suppressed this complication in the main text.

19. For a partition containing only two members (call them C_1 and C_2), the unconditional credence form of the Law of Total Probability tells us that

$$cr(A) = cr(A \mid C_1) \cdot cr(C_1) + cr(A \mid C_2) \cdot cr(C_2)$$

The conditional credence form (generated by the procedure described in Section 3.1.2) tells us that for any E with $cr(E) > 0$,

$$cr(A \mid E) = cr(A \mid C_1 \mathbin{\&} E) \cdot cr(C_1 \mid E) + cr(A \mid C_2 \mathbin{\&} E) \cdot cr(C_2 \mid E)$$

20. One caveat I *won't* get into is that Lewis's original (1980) formulation of the Principal Principle becomes inconsistent if we allow propositions about chances to have chances of their own, and those chances of chances may be nonextreme. For why we might allow this, and how Lewis (and others) reformulated the Principal Principle in response, see Lewis (1994) and the literature that followed.

21. Equation (5.9) directs the assignment of your unconditional credences only when information about the opinion of a particular expert is your *total* relevant evidence concerning proposition A. If you have additional information about A (perhaps the opinion of a second expert?), the relevant condition in the conditional credence on the left-hand side of Equation (5.9) is no longer just $cr_E(A) = x$. (See Exercise (5.2) for more on this point.)

22. Supposing that your future credences result from your present credences by conditionalization guarantees that your future self will possess at least as much evidence as your present self. But it also has the advantage of guaranteeing that future and present selves both work from the same hypothetical prior distribution (because of the Hypothetical Priors Theorem, Section 4.3). It's worth thinking about whether an agent should defer to the opinions of a database expert who, while having evidence that's a strict superset of the agent's, analyzes that evidence using different epistemic standards.

23. The justification I've just provided for Equation (5.11) explicitly uses every one of the enumerated conditions except Condition 3. Condition 3 is necessary so that the conditional credence in Equation (5.11) is well defined according to the Ratio Formula.

24. One complication here is that van Fraassen sometimes describes Reflection as relating attitudes, but at other times portrays it as being about various *acts* of commitment, and therefore more directly concerned with assertions and avowals than with particular mental states.

25. The Reflection Principle applies to times t_i and t_j with j strictly greater than i. What would happen if we applied it when $j = i$? In that case we'd have a principle for how an agent's current credences should line up with her credences about her current credences. This principle would engage the results of an agent's introspection to determine what her current credences are. An agent's credences about her own current credences are her **higher-order credences**, and they have been the subject of much Bayesian scrutiny (e.g., Skyrms 1980b). The core issue is how much access a rational agent is required to have to the contents of her own mind.

26. Joyce (2005) reports that this sort of problem was first identified by John Venn in the 1800s.

27. This example is adapted from one in Salmon (1966, pp. 66–7). A related example is van Fraassen's (1989) Cube Factory, which describes a factory making cubes of various sizes and asks how confident I should be that a given manufactured cube has a size falling within a particular range. The Principle of Indifference yields conflicting answers depending on whether cube size is characterized using side length, face area, or volume.

28. In Chapter 14 we will discuss other potential responses to this kind of ignorance.

29. What about cases in which an agent *has* ruled out the proposition Q? Should rational agents assign credences conditional on conditions that they've ruled out? For discussion and references on this question, see Titelbaum (2013a, Ch. 5).

30. I was careful to define the Ratio Formula so that it simply goes silent when $cr(Q) = 0$, and is therefore in need of *supplementation* if we want to constrain values like $cr(2 \mid 2)$. Other authors define the Ratio Formula so that it contains the same equation as ours but leaves off the restriction to $cr(Q) > 0$ cases. This forces an impossible calculation when $cr(Q) = 0$. Alternatively, one can leave the Ratio Formula unrestricted but make its equation $cr(P \mid Q) \cdot cr(Q) = cr(P \& Q)$. This has the advantage of being *true* even when $cr(Q) = 0$ (because $cr(P \& Q)$ will presumably equal 0 as well), but does no better than our Ratio Formula in constraining the value of $cr(2 \mid 2)$. (Any value we fill in for that conditional credence will make the relevant product-equation true.)

31. For a historical overview of the approach and detailed comparison of the disparate formal systems, see Makinson (2011).

32. Seidenfeld, Schervish, and Kadane (2017) shows that this pattern generalizes: At each infinite cardinality, we cannot secure the relevant Conglomerability principle with Additivity principles of lower cardinalities; Conglomerability at a particular level requires Additivity at that same level.

33. I got the example that follows from Brian Weatherson.

34. Contrast our move from comparative to quantitative representations of doxastic attitudes in Chapter 1. There the additional structure of a numerical representation allowed us to model features like confidence-gap sizes, which plausibly make a difference to agents' real-world decisions.

35. Let me quickly tie up one loose end: This section discussed cases in which it might be rational for an agent to assign unconditional credence 0 to a proposition without ruling it out. All the cases in which this might be rational involve credence assignments over infinite partitions. For the rest of this book we will be working with finite partitions, and will revert to the assumption we were making prior to this section that credence 0 always represents ruling something out.

36. Actually, Jeffrey's original proposal was a bit more complicated than that. In Jeffrey (1965) he began with a set of propositions $\{B_1, B_2, \ldots, B_n\}$ in which the credence change originated, but did not require the B_m to form a partition. Instead, he constructed a set of "atoms", which we can think of as state-descriptions constructed from the B_m. (Each atom was a consistent conjunction in which each B_m appeared exactly once, either affirmed or negated.) The Rigidity condition (which Jeffrey sometimes called "invariance") and Jeffrey Conditionalization were then applied to these atoms rather than directly to the B_m in which the credence change originated.

Notice that in this construction the atoms form a partition. Further, Jeffrey recognized that if the B_m themselves formed a partition, the atoms wound up in a one-to-one correspondence with the B_m to which they were logically equivalent. I think it's for this reason that Jeffrey later (2004, Ch. 3) dropped the business with "atoms" and applied his probability kinematics directly to any finite partition.

37. Though see Lange (2000) for an argument that this order-dependence is not a problem because the character of the experiences changes when they're temporally rearranged.

38. Interestingly, the main thrust of van Fraassen's article is that while Maximum Entropy *is* capable of providing a solution to the Judy Benjamin Problem, that solution is intuitively unappealing.

39. Because we're going to be using the entropy values only for comparative purposes, in the end it won't make a difference what base we use for the logarithms in Equation (5.13). But just to make your answers easily checkable with others', please follow Jaynes (1957a) in using the natural log ln.

40. I owe this problem to Sarah Moss.

Glossary for Volume 1

actual world The possible world in which we live. Events that actually happen happen in the actual world. 26

admissible evidence Evidence that, if it has any effect on an agent's credence in an outcome of an event, does so by way of affecting the agent's credences about the outcome's objective chance. 135

analyst expert Expert to whom one defers because of her skill at forming attitudes on the basis of evidence. 142

antecedent In a conditional of the form "If P, then Q," P is the antecedent. 26

atomic proposition A proposition in language $\mathcal{L}$ that does not contain any connectives or quantifiers. An atomic proposition is usually represented either as a single capital letter (P, Q, R, etc.) or as a predicate applied to some constants (Fa, Lab, etc.). 26

Base Rate Fallacy Assigning a posterior credence to a hypothesis that overemphasizes the likelihoods associated with one's evidence and underemphasizes one's prior in the hypothesis. 97

Bayes factor For a given piece of evidence, the ratio of the likelihood of the hypothesis to the likelihood of the catchall. An update by Conditionalization multiplies your odds for the hypothesis by the Bayes factor of your evidence. 98

Bayes Net A diagram of causal relations among variables developed from information about probabilistic dependencies among them. 75

Bayes's Theorem For any H and E in $\mathcal{L}$, $\operatorname{cr}(H \mid E) = \operatorname{cr}(E \mid H) \cdot \operatorname{cr}(H)/\operatorname{cr}(E)$. 61

Belief Closure If some subset of the propositions an agent believes entails a further proposition, rationality requires the agent to believe that further proposition as well. 6

Belief Consistency Rationality requires the set of propositions an agent believes to be logically consistent. 6

Bertrand's Paradox When asked how probable it is that a chord of a circle is longer than the side of an inscribed equilateral triangle, the Principle of Indifference produces different answers depending on how the chord is specified. 147

catchall The proposition that the hypothesis H under consideration is false (in other words, the proposition $\sim H$). 64

classical probability The number of favorable event outcomes divided by the total number of outcomes possible. 124

classificatory concept Places an entity in one of a small number of kinds. 4

common cause A single event that causally influences at least two other events. 73

commutative Updating by Conditionalization is commutative in the sense that updating first on E then on E' has the same effect as updating in the opposite order. 94

comparative concept Places one entity in order with respect to another. 4

Comparative Entailment For propositions P and Q, if $P \vDash Q$ then rationality requires an agent to be at least as confident of Q as P. 11

condition In a conditional credence, the proposition the agent supposes. 56

conditional credence A degree of belief assigned to an ordered pair of propositions, indicating how confident the agent is that the first proposition is true on the supposition that the second is. 56

conditional independence When $cr(Q \& R) > 0$, P is probabilistically independent of Q conditional on R just in case $cr(P \mid Q \& R) = cr(P \mid R)$. 68

Conditionalization For any time t_i and later time t_j, if proposition E in $\mathcal{L}$ represents everything the agent learns between t_i and t_j, and $cr_i(E) > 0$, then for any H in $\mathcal{L}$, $cr_j(H) = cr_i(H \mid E)$ (Bayesians' traditional updating rule). 91

Conglomerability For each proposition P and partition $\{Q_1, Q_2, Q_3, \ldots\}$ in $\mathcal{L}$, $cr(P)$ is no greater than the largest $cr(P \mid Q_i)$ and no less than the least $cr(P \mid Q_i)$. 153

conjunction $P \& Q$ is a conjunction; P and Q are its conjuncts. 26

Conjunction Fallacy Being more confident in a conjunction than you are in one of its conjuncts. 39

consequent In a conditional of the form "If P, then Q," Q is the consequent. 26

consistent The propositions in a set are consistent when at least one possible world makes all the propositions true. 28

constant A lower-case letter in language $\mathcal{L}$ representing an object in the universe of discourse. 30

contingent A proposition that is neither a tautology nor a contradiction. 28

contradiction A proposition that is false in every possible world. 28

Contradiction For any contradiction F in $\mathcal{L}$, $cr(F) = 0$. 34

Countable Additivity For any countable set $\{Q_1, Q_2, Q_3, \ldots\}$ of mutually exclusive propositions in $\mathcal{L}$, $cr(Q_1 \lor Q_2 \lor Q_3 \lor \ldots) = cr(Q_1) + cr(Q_2) + cr(Q_3) + \ldots$. 152

credence Degree of belief. 3

cumulative Updating by Conditionalization is cumulative in the sense that updating first on evidence E and then on evidence E' has the same net effect as updating once, on the conjunction E & E'. 94

database expert Expert to whom one defers because her evidence includes one's own, and more. 142

Decomposition For any propositions P and Q in $\mathcal{L}$, $\mathrm{cr}(P)$ = $\mathrm{cr}(P$ & $Q)$ + $\mathrm{cr}(P$ & $\sim Q)$. 34

deference principle Any principle directing an agent to align her current credences with some other distribution (such as objective chances, credences of an expert, or credences of her future self). 141

direct inference Determining how likely one is to obtain a particular experimental result from probabilistic hypotheses about the setup. 61

disjunction $P \vee Q$ is a disjunction; P and Q are its disjuncts. 26

disjunctive normal form The disjunctive normal form of a non-contradictory proposition is the disjunction of state-descriptions that is equivalent to that proposition. 30

distribution An assignment of real numbers to each proposition in language $\mathcal{L}$. 31

doxastic attitude A belief-like representational propositional attitude. 3

doxastically possible worlds The subset of possible worlds that a given agent entertains. 37

entailment P entails Q ($P \vDash Q$) just in case there is no possible world in which P is true and Q is false. On a Venn diagram, the P-region is wholly contained in the Q-region. 28

Entailment For any propositions P and Q in $\mathcal{L}$, if $P \vDash Q$ then $\mathrm{cr}(P) \leq \mathrm{cr}(Q)$. 34

epistemic standards Applying an agent's ultimate epistemic standards to her total evidence at a given time yields her doxastic attitudes at that time. Bayesians represent ultimate epistemic standards as hypothetical priors. 107

Equivalence For any propositions P and Q in $\mathcal{L}$, if $P \; \dashv\vDash \; Q$ then $\mathrm{cr}(P)$ = $\mathrm{cr}(Q)$. 34

equivalent Equivalent propositions are associated with the same set of possible worlds. 28

evidential probability The degree to which a body of evidence probabilifies a hypothesis, understood as independent of any particular agent's attitudes. 131

evidentialism The position that what attitudes are rationally permissible for an agent supervenes on her evidence. 130

exhaustive The propositions in a set are jointly exhaustive if each possible world makes at least one of the propositions in the set true. 28

falsification A piece of evidence falsifies a hypothesis if it refutes that hypothesis relative to one's background assumptions. 67

Finite Additivity For any mutually exclusive propositions P and Q in $\mathcal{L}$, $cr(P \vee Q) = cr(P) + cr(Q)$ (one of the three probability axioms). 32

Finite Additivity (Extended) For any finite set of mutually exclusive propositions $\{P_1, P_2, \ldots, P_n\}$, $cr(P_1 \vee P_2 \vee \ldots \vee P_n) = cr(P_1) + cr(P_2) + \ldots + cr(P_n)$. 34

frequency theory An interpretation of probability according to which the probability is x that event A will have outcome B just in case fraction x of events like A have outcomes like B. 125

Gambler's Fallacy Expecting later outcomes of an experiment to "compensate" for unexpected previous results despite the probabilistic independence of future results from those in the past. 68

General Additivity For any propositions P and Q in $\mathcal{L}$, $cr(P \vee Q) = cr(P) + cr(Q) - cr(P \,\&\, Q)$. 34

higher-order credences An agent's credences about her own current credences. Includes both her credences about what her current credence-values *are* and her credences about what those values *should be*. 171

Humphreys's Paradox Difficulty for the propensity interpretation of probability that when the probability of E given H can be understood in terms of propensities it is often difficult to interpret the probability of H given E as a propensity as well. 168

hypothetical frequency theory Interpretation of probability that looks not at the proportion of actual events producing a particular outcome but instead at the proportion of such events that would produce that outcome in the limit. 127

hypothetical prior distribution A regular, probabilistic distribution used to represent an agent's ultimate epistemic standards. The agent's credence distribution at a given time can be recovered by conditionalizing her hypothetical prior on her total evidence at that time. 110

Hypothetical Priors Theorem Given any finite series of credence distributions $\{cr_1, cr_2, \ldots, cr_n\}$ each of which satisfies the probability axioms and Ratio Formula, let E_i be a conjunction of the agent's total evidence at t_i. If each cr_i is related to cr_{i+1} as specified by Conditionalization, then there exists at least one regular probability distribution Pr_H such that for all $1 \leq i \leq n$, $cr_i(\cdot) = Pr_H(\cdot \,|\, E_i)$. 110

IID trials Independent, identically distributed probabilistic events. Trials are IID if the probabilities associated with a given trial are unaffected by the outcomes of other trials (independence), and if each trial has the same probability of producing a given outcome (identically distributed). 87

inconsistent The propositions in a set are inconsistent when there is no possible world in which all of them are true. 28

independence When $cr(Q) > 0$, proposition P is probabilistically independent of proposition Q relative to cr just in case $cr(P\,|\,Q) = cr(P)$. 65

infinitesimal A number that is greater than zero but less than any positive real number. 155

initial prior distribution Credence distribution assigned by an agent before she possessed any contingent evidence. 106

interpretations of probability Philosophical theories about the nature of probability and the meanings of linguistic probability expressions. 124

inverse inference Determining how likely a probabilistic hypothesis is on the basis of a particular run of experimental data. 62

irrelevant Probabilistically independent. 65

Jeffrey Conditionalization Proposed by Richard C. Jeffrey as an alternative updating rule to Conditionalization, holds that for any t_i and t_j with $i < j$, any A in $\mathcal{L}$, and a finite partition $\{B_1, B_2, \ldots, B_n\}$ in $\mathcal{L}$ whose members each have nonzero cr_i, $cr_j(A) = cr_i(A\,|\,B_1) \cdot cr_j(B_1) + cr_i(A\,|\,B_2) \cdot cr_j(B_2) + \ldots + cr_i(A\,|\,B_n) \cdot cr_j(B_n)$. 157

Judy Benjamin Problem An example proposed by Bas van Fraassen in which an agent's experience directly alters some of her conditional credence values. van Fraassen argued that this example could not be addressed by traditional Conditionalization or by Jeffrey Conditionalization. 161

just in case If and only if. 26

Kolmogorov's axioms The three axioms (Non-Negativity, Normality, and Finite Additivity) that provide necessary and sufficient conditions for a probability distribution. 32

Law of Total Probability For any proposition P and finite partition $\{Q_1, Q_2, \ldots, Q_n\}$ in $\mathcal{L}$, $cr(P) = cr(P\,|\,Q_1) \cdot cr(Q_1) + cr(P\,|\,Q_2) \cdot cr(Q_2) + \ldots + cr(P\,|\,Q_n) \cdot cr(Q_n)$. 59

likelihood The probability of some particular piece of evidence on the supposition of a particular hypothesis—$cr(E\,|\,H)$. 62

linear combination When applied to three numerical variables, z is a linear combination of x and y just in case there exist constants a and b such that $z = ax + by$. For instance, Finite Additivity requires $cr(X \vee Y) = cr(X) + cr(Y)$, making the credence in a disjunction a linear combination of the credences in its mutually exclusive disjuncts (with the two constants a and b each set to 1). When applied to probability distributions, distribution Pr_z over language $\mathcal{L}$ is a linear combination of distributions Pr_x and Pr_y over $\mathcal{L}$ just in case there exists some $0 \leq \alpha \leq 1$ such that for every proposition $P \in \mathcal{L}$, $Pr_z(P) = \alpha \cdot Pr_x(P) + (1 - \alpha) \cdot Pr_y(P)$. 54

Lockean thesis Connects believing a proposition with having a degree of confidence in that proposition above a numerical threshold. 15

logical probability The degree to which a body of evidence probabilifies a hypothesis, understood as a logical relation similar to deductive entailment. 131

logically possible A world is logically possible when it violates no laws of logic. So, for instance, a world with no gravity is physically impossible because it violates the laws of physics, but it is logically possible. 37

Lottery Paradox Paradox for requirements of logical belief consistency and closure involving a lottery with a large number of tickets. 8

material biconditional A material biconditional $P \equiv Q$ is true just in case P and Q are both true or P and Q are both false. 26

material conditional A material conditional $P \supset Q$ is false just in case its antecedent P is true and its consequent Q is false. 26

Maximality For any proposition P in $\mathcal{L}$, $\mathrm{cr}(P) \leqslant 1$. 34

Maximum Entropy Principle Given any partition of the space of possibilities, and any set of constraints on allowable credence distributions over that partition, the Maximum Entropy Principle selects the allowable distribution with the highest entropy. 147

monotonicity If a property is monotonic, then whenever proposition E bears the property, every conjunction containing E as a conjunct bears the property as well. 101

Monty Hall Problem A famous probabilistic puzzle case, demonstrating the importance of taking an agent's total evidence into account. 102

Multiplication When P and Q have nonextreme cr-values, P and Q are probabilistically independent relative to cr if and only if $\mathrm{cr}(P \& Q) = \mathrm{cr}(P) \cdot \mathrm{cr}(Q)$. 65

mutually exclusive The propositions in a set are mutually exclusive when there is no possible world in which more than one of the propositions is true. 28

negation $\sim P$ is the negation of P. 26

Negation For any proposition P in $\mathcal{L}$, $\mathrm{cr}(\sim P) = 1 - \mathrm{cr}(P)$. 33

negative relevance When $\mathrm{cr}(Q) > 0$, Q is negatively relevant to P relative to cr just in case $\mathrm{cr}(P \mid Q) < \mathrm{cr}(P)$. 66

Non-Negativity For any proposition P in $\mathcal{L}$, $\mathrm{cr}(P) \geq 0$ (one of the three probability axioms). 32

nonmonotonicity Probabilistic relations are nonmonotonic in the sense that even if H is highly probable given E, H might be improbable given the conjunction of E with some E'. 102

Normality For any tautology T in $\mathcal{L}$, cr(T) = 1 (one of the three probability axioms). 32

normalization factor In an update by Conditionalization, state-descriptions inconsistent with E (the evidence learned) have their unconditional credences sent to zero. The remaining state-descriptions all have their unconditional credences multiplied by the same normalization factor, equal to the reciprocal of E's prior. 119

normative distinction The normative distinction between Subjective and Objective Bayesians concerns the strength of rationality's requirements. Distinguished this way, Objective Bayesians hold that there is exactly one rationally permissible set of epistemic standards (/hypothetical priors), so that any body of total evidence gives rise to a unique rational attitude toward any particular proposition. Subjective Bayesians deny that rational requirements are strong enough to mandate a unique attitude in every case. 129

objective chance A type of physical probability that can be applied to the single case. 128

observation selection effect Effect on the appropriate conclusions to draw from a piece of evidence introduced by the manner in which that evidence was obtained (for example, the method by which a sample was drawn). 102

odds If an agent's unconditional credence in P is cr(P), her odds for P are cr(P) : cr($\sim P$), and her odds against P are cr($\sim P$) : cr(P). 45

partition A mutually exclusive, jointly exhaustive set of propositions. On a Venn diagram, the regions representing propositions in a partition combine to fill the entire rectangle without overlapping at any point. 29

Partition For any finite partition of propositions in $\mathcal{L}$, the sum of their unconditional cr-values is 1. 34

permissive case An example in which two agents with identical total evidence assign different credences without either agent's thereby being irrational. Objective Bayesians in the normative sense deny the existence of permissive cases. 131

positive relevance When cr(Q) > 0, Q is positively relevant to P relative to cr just in case cr($P\,|\,Q$) > cr(P). 66

possible worlds Different ways the world might have come out. Possible worlds are maximally specified—for any event and any possible world that event either does or does not occur in that world—and the possible worlds are plentiful enough such that for any combination of events that *could* happen, there is a possible world in which that combination of events *does* happen. 26

posterior The probability of some hypothesis on the supposition of a particular piece of evidence—$P(H\,|\,E)$. 63

practical rationality Concerns the connections between attitudes and actions. 7

predicate A capital letter representing a property or relation in language $\mathcal{L}$. 30

Preface Paradox Paradox for requirements of logical belief consistency and closure in which the preface to a nonfiction book asserts that at least one of the claims in the book is false. 8

Principal Principle David Lewis's proposal for how rational credences concerning an event incorporate suppositions about the objective chances of that event's possible outcomes. 136

Principle of Indifference If an agent has no evidence favoring any proposition in a partition over any other, she should spread her credence equally over the members of the partition. 146

Principle of the Common Cause When event outcomes are probabilistically correlated, either one causes the other or they have a common cause. 73

Principle of Total Evidence A rational agent's credence distribution takes into account all of the evidence she possesses. 102

prior An unconditional probability; the probability of a proposition before anything has been supposed. For example, an agent's prior credence in a particular hypothesis H is $cr(H)$. 63

probabilism The thesis that rationality requires an agent's credences to satisfy the probability axioms. 33

probabilistically independent When $cr(Q) > 0$, P is probabilistically independent of Q relative to cr just in case $cr(P \mid Q) = cr(P)$. 65

probability axioms Kolmogorov's axioms. 32

probability distribution Any distribution satisfying Kolmogorov's probability axioms. 32

probability kinematics What Richard C. Jeffrey, its inventor, called the updating rule now generally known as "Jeffrey Conditionalization". 157

probability table A table that assigns unconditional credences to each member in a partition. To satisfy the probability axioms, the values in each row must be nonnegative and all the values must sum to 1. When the partition members are state-descriptions of a language $\mathcal{L}$, the values in the probability table suffice to specify all of the agent's credences over $\mathcal{L}$. 41

problem of the single case The challenge of interpreting probability such that single (and perhaps non-repeatable) events may receive nonextreme probabilities. 127

propensity theory Interpretation of probability identifying probability with a physical arrangement's quantifiable tendency to produce outcomes of a particular kind. 127

proposition An abstract entity expressible by a declarative sentence and capable of having a truth-value. 3

propositional attitude An attitude adopted by an agent toward a proposition or set of propositions. 3

propositional connective One of five truth-functional symbols ($\sim$, &, $\vee$, $\supset$, $\equiv$) used to construct larger propositions from atomic propositions. 26

quantitative concept Characterizes an entity by ascribing it a numerical value. 4

Ratio Formula For any P and Q in $\mathcal{L}$, if $\mathrm{cr}(Q) > 0$ then $\mathrm{cr}(P \,|\, Q) = \mathrm{cr}(P \,\&\, Q)/\mathrm{cr}(Q)$. The Bayesian rational constraint relating an agent's conditional credences to her unconditional credences. 57

reference class problem When considering a particular event and one of its possible outcomes, the frequency with which this type of event produces that type of outcome depends on which reference class (event-type) we choose out of the many to which the event belongs. 126

Reflection Principle For any proposition A in $\mathcal{L}$, real number x, and times t_i and t_j with $j > i$, rationality requires $\mathrm{cr}_i(A \,|\, \mathrm{cr}_j(A) = x) = x$. 143

refutation P refutes Q just in case P entails $\sim Q$. When P refutes Q, every world that makes P true makes Q false. 28

regular A distribution that does not assign the value 0 to any logically contingent propositions. 99

Regularity Principle In a rational credence distribution, no logically contingent proposition receives unconditional credence 0. 99

relevant Not probabilistically independent. 66

Rigidity condition For any A in $\mathcal{L}$ and any B_m in the finite partition $\{B_1, B_2, \ldots, B_n\}$, $\mathrm{cr}_j(A \,|\, B_m) = \mathrm{cr}_i(A \,|\, B_m)$. This condition obtains between t_i and t_j just in case the agent Jeffrey Conditionalizes across $\{B_1, B_2, \ldots, B_n\}$. 159

rule of succession Laplace's rule directing an agent who has witnessed h heads on n independent flips of a coin to set credence $(h + 1)/(n + 2)$ that the next flip will come up heads. 169

screening off R screens off P from Q when P is unconditionally relevant to Q but not relevant to Q conditional on either R or $\sim R$. 68

semantic distinction When classified according to the semantic distinction, Subjective Bayesians take "probability" talk to reveal the credences of agents, while Objective Bayesians assign "probability" assertions truth-conditions independent of the attitudes of particular agents or groups of agents. 129

sigma algebra A set of sets closed under union, intersection, and complementation. A probability distribution can be assigned over a sigma algebra containing sets of possible worlds instead of over a language containing propositions. 52

Simple Binarist A made-up character who describes agents' doxastic propositional attitudes exclusively in terms of belief, disbelief, and suspension of judgment. 5

Simpson's Paradox Two propositions may be correlated conditional on each member of a partition yet anti-correlated unconditionally. 70

state-description A conjunction of language $\mathcal{L}$ in which (1) each conjunct is either an atomic proposition of $\mathcal{L}$ or its negation; and (2) each atomic proposition of $\mathcal{L}$ appears exactly once. 29

straight rule Reichenbach's name for the norm setting an agent's credence that the next event of type A will produce an outcome of type B exactly equal to the observed frequency of B-outcomes in past A-events. 169

strict Conditionalization Another name for the Conditionalization updating rule. The "strict" is usually used to emphasize a contrast with Jeffrey Conditionalization. 160

subadditive In a subadditive distribution, there exist mutually exclusive P and Q in $\mathcal{L}$ such that $\mathrm{cr}(P \vee Q) < \mathrm{cr}(P) + \mathrm{cr}(Q)$. 51

substitution instance A substitution instance of a quantified sentence is produced by removing the quantifier and replacing its variable throughout what remains with the same constant. 31

superadditive In a superadditive distribution, there exist mutually exclusive P and Q in $\mathcal{L}$ such that $\mathrm{cr}(P \vee Q) > \mathrm{cr}(P) + \mathrm{cr}(Q)$. 47

supervenience A-properties supervene on B-properties just in case any two objects that differ in their A-properties also differ in their B-properties. For example, one's score on a test supervenes on the answers one provides; if two students got different scores on the same test, their answers must have differed. 58

tautology A proposition that is true in every possible world. 28

theoretical rationality Evaluates representational attitudes in their capacity as representations, without considering how they influence action. 7

truth-value *True* and *false* are truth-values. We assume propositions are capable of having truth-values. 3

unconditional credence An agent's degree of belief in a proposition, without making any suppositions beyond her current background information. 32

Uniqueness Thesis Given any proposition and body of total evidence, there is exactly one attitude it is rationally permissible for agents with that body of total evidence to adopt toward that proposition. 129

universe of discourse The set of objects under discussion. 30

ur-prior Alternate name for a hypothetical prior distribution. 106

Venn Diagram Diagram in which an agent's doxastically possible worlds are represented as points in a rectangle. Propositions are represented by regions containing those points, with the area of a region often representing the agent's credence in an associated proposition. 26

Bibliography of Volume 1

Adams, Ernest (1965). The Logic of Conditionals. *Inquiry* 8, pp. 166–97.

Alchourrón, Carlos E., Peter Gärdenfors, and David Makinson (1985). On the Logic of Theory Change: Partial Meet Contraction and Revision Functions. *The Journal of Symbolic Logic* 50, pp. 510–30.

Arntzenius, Frank (1993). The Common Cause Principle. *PSA: Proceedings of the Biennial Meeting of the Philosophy of Science Association* 2, pp. 227–37.

Bartha, Paul and Christopher R. Hitchcock (1999). No One Knows the Date or the Hour: An Unorthodox Application of Rev. Bayes's Theorem. *Philosophy of Science* 66, S339–53.

Bergmann, Merrie, James Moor, and Jack Nelson (2013). *The Logic Book*. 6th edition. New York: McGraw Hill.

Bernoulli, Jacob (1713). *Ars Conjectandi*. Basiliae.

Bertrand, Joseph (1888/1972). *Calcul des probabilités*. 2nd edition. New York: Chelsea Publishing Company.

Bickel, P.J., E.A. Hammel, and J.W. O'Connell (1975). Sex Bias in Graduate Admissions: Data from Berkeley. *Science* 187, pp. 398–404.

Bolzano, Bernard (1837/1973). *Wissenschaftslehre*. Translated by Jan Berg under the title *Theory of Science*. Dordrecht: Reidel.

Bradley, Darren (2010). Conditionalization and Belief *De Se*. *Dialectica* 64, pp. 247–50.

Bradley, Darren (2015). *A Criticial Introduction to Formal Epistemology*. London: Bloomsbury.

Carnap, Rudolf (1945). On Inductive Logic. *Philosophy of Science* 12, pp. 72–97.

Carnap, Rudolf (1947). On the Application of Inductive Logic. *Philosophy and Phenomenological Research* 8, pp. 133–48.

Carnap, Rudolf (1950). *Logical Foundations of Probability*. Chicago: University of Chicago Press.

Carnap, Rudolf (1962b). The Aim of Inductive Logic. In: *Logic, Methodology, and the Philosophy of Science*. Ed. by P. Suppes, E. Nagel, and A. Tarski. Stanford University: Stanford University Press, pp. 303–18.

Cartwright, Nancy (1979). Causal Laws and Effective Strategies. *Noûs* 13, pp. 419–37.

Christensen, David (2004). *Putting Logic in its Place*. Oxford: Oxford University Press.

Davidson, Donald (1984). *Inquiries into Truth and Interpretation*. Oxford: Clarendon Press.

de Finetti, Bruno (1931/1989). Probabilism: A Critical Essay on the Theory of Probability and the Value of Science. *Erkenntnis* 31, pp. 169–223. Translation of B. de Finetti, *Probabilismo*, Logos 14, pp. 163–219.

de Finetti, Bruno (1995). *Filosofia della probabilità*. Ed. by Alberto Mura. Milan: Il Saggiatore.

Earman, John (1992). *Bayes or Bust? A Critical Examination of Bayesian Confirmation Theory*. Cambridge, MA: The MIT Press.

Easwaran, Kenny (2014b). Regularity and Hyperreal Credences. *Philosophical Review* 123, pp. 1–41.

Eddington, A. (1939). *The Philosophy of Physical Science*. Cambridge: Cambridge University Press.

Elga, Adam (2007). Reflection and Disagreement. *Noûs* 41, pp. 478–502.

Ellenberg, Jordan (2014). *How Not to Be Wrong: The Power of Mathematical Thinking*. New York: Penguin Press.

Ellis, Robert Leslie (1849). On the Foundations of the Theory of Probabilities. *Transactions of the Cambridge Philosophical Society* VIII, pp. 1–6.

Feldman, Richard (2007). Reasonable Religious Disagreements. In: *Philosophers without Gods: Meditations on Atheism and the Secular Life*. Ed. by Louise M. Antony. Oxford: Oxford University Press.

Fitelson, Branden (2008). A Decision Procedure for Probability Calculus with Applications. *The Review of Symbolic Logic* 1, pp. 111–25.

Fitelson, Branden (2015). The Strongest Possible Lewisian Triviality Result. *Thought* 4, pp. 69–74.

Fitelson, Branden and Alan Hájek (2014). Declarations of Independence. *Synthese* 194, pp. 3979–95.

Foley, Richard (1987). *The Theory of Epistemic Rationality*. Cambridge, MA: Harvard University Press.

Foley, Richard (1993). *Working without a Net*. Oxford: Oxford University Press.

Foley, Richard (2009). Beliefs, Degrees of Belief, and the Lockean Thesis. In: *Degrees of Belief*. Ed. by Franz Huber and Christoph Schmidt-Petri. Vol. 342. Synthese Library. Springer, pp. 37–48.

Galavotti, Maria Carla (2005). *Philosophical Introduction to Probability*. CSLI Lecture Notes 167. Stanford, CA: CSLI Publications.

Gillies, Donald (2000). Varieties of Propensity. *British Journal for the Philosophy of Science* 51, pp. 807–35.

Good, I.J. (1968). The White Shoe qua Herring Is Pink. *British Journal for the Philosophy of Science* 19, pp. 156–7.

Good, I.J. (1971). Letter to the Editor. *The American Statistician* 25, pp. 62–3.

Hacking, Ian (1971). The Leibniz-Carnap Program for Inductive Logic. *The Journal of Philosophy* 68, pp. 597–610.

Hacking, Ian (2001). *An Introduction to Probability and Inductive Logic*. Cambridge: Cambridge University Press.

Hájek, Alan (1996). 'Mises Redux'—Redux: Fifteen Arguments against Finite Frequentism. *Erkenntnis* 45, pp. 209–27.

Hájek, Alan (2003). What Conditional Probability Could Not Be. *Synthese* 137, pp. 273–323.

Hájek, Alan (2009b). Fifteen Arguments against Hypothetical Frequentism. *Erkenntnis* 70, pp. 211–35.

Hájek, Alan (2011a). Conditional Probability. In: *Philosophy of Statistics*. Ed. by Prasanta S. Bandyopadhyay and Malcolm R. Forster. Vol. 7. Handbook of the Philosophy of Science. Amsterdam: Elsevier, pp. 99–136.

Hájek, Alan (2011b). Triviality Pursuit. *Topoi* 30, pp. 3–15.

Hájek, Alan (2019). Interpretations of Probability. In: *The Stanford Encyclopedia of Philosophy*. Ed. by Edward N. Zalta. Fall 2019. URL: http://plato.stanford.edu/archives/fall2019/entries/probability-interpret/.

Hall, Ned (2004). Two Mistakes about Credence and Chance. *Australasian Journal of Philosophy* 82, pp. 93–111.

Hart, Casey and Michael G. Titelbaum (2015). Intuitive Dilation? *Thought* 4, pp. 252–62.

Hitchcock, Christopher R. (2021). Probabilistic Causation. In: *The Stanford Encyclopedia of Philosophy*. Ed. by Edward N. Zalta. Spring 2021. URL: https://plato.stanford.edu/archives/spr2021/entries/causation-probabilistic/.

Holton, Richard (2014). Intention as a Model for Belief. In: *Rational and Social Agency: The Philosophy of Michael Bratman*. Ed. by Manuel Vargas and Gideon Yaffe. Oxford: Oxford University Press, pp. 12–37.

Hosiasson-Lindenbaum, Janina (1940). On Confirmation. *Journal of Symbolic Logic* 5, pp. 133–48.

Howson, Colin (2014). Finite Additivity, Another Lottery Paradox and Conditionalisation. *Synthese* 191, pp. 989–1012.

Howson, Colin and Peter Urbach (2006). *Scientific Reasoning: The Bayesian Approach*. 3rd edition. Chicago: Open Court.

Hume, David (1739–40/1978). *A Treatise of Human Nature*. Ed. by L.A. Selby-Bigge and Peter H. Nidditch. 2nd edition. Oxford: Oxford University Press.

Humphreys, Paul (1985). Why Propensities Cannot Be Probabilities. *Philosophical Review* 94, pp. 557–70.

Jackson, Elizabeth G. (2020). The Relationship between Belief and Credence. *Philosophy Compass* 15, pp. 1–13.

Jaynes, E.T. (1957a). Information Theory and Statistical Mechanics I. *Physical Review* 106, pp. 62–30.

Jaynes, E.T. (1957b). Information Theory and Statistical Mechanics II. *Physical Review* 108, pp. 171–90.

Jeffrey, Richard C. (1965). *The Logic of Decision*. 1st edition. McGraw-Hill Series in Probability and Statistics. New York: McGraw-Hill.

Jeffrey, Richard C. (1970). Dracula Meets Wolfman: Acceptance vs. Partial Belief. In: *Induction, Acceptance, and Rational Belief*. Ed. by M. Swain. Dordrecht: Reidel, pp. 157–85.

Jeffrey, Richard C. (2004). *Subjective Probability: The Real Thing*. Cambridge: Cambridge University Press.

Joyce, James M. (1999). *The Foundations of Causal Decision Theory*. Cambridge: Cambridge University Press.

Joyce, James M. (2005). How Probabilities Reflect Evidence. *Philosophical Perspectives* 19, pp. 153–78.

Keynes, John Maynard (1921). *Treatise on Probability*. London: Macmillan and Co., Limited.

Kim, Jaegwon (1988). What Is "Naturalized Epistemology"? *Philosophical Perspectives* 2, pp. 381–405.

Kolmogorov, A.N. (1933/1950). *Foundations of the Theory of Probability*. Translation edited by Nathan Morrison. New York: Chelsea Publishing Company.

Kornblith, Hilary (1993). Epistemic Normativity. *Synthese* 94, pp. 357–76.

Kuhn, Thomas S. (1957). *The Copernican Revolution: Planetary Astronomy in the Development of Western Thought*. New York: MJF Books.

Kyburg Jr, Henry E. (1961). *Probability and the Logic of Rational Belief*. Middletown: Wesleyan University Press.

Kyburg Jr, Henry E. (1970). Conjunctivitis. In: *Induction, Acceptance, and Rational Belief*. Ed. by M. Swain. Boston: Reidel, pp. 55–82.

Lance, Mark Norris (1995). Subjective Probability and Acceptance. *Philosophical Studies* 77, pp. 147–79.

Lange, Alexandra (2019). Can Data Be Human? The Work of Giorgia Lupi. *The New Yorker*. Published May 25, 2019.

Lange, Marc (2000). Is Jeffrey Conditionalization Defective by Virtue of Being Non-commutative? Remarks on the Sameness of Sensory Experience. *Synthese* 123, pp. 393–403.

Laplace, Pierre-Simon (1814/1995). *Philosophical Essay on Probabilities*. Translated from the French by Andrew Dale. New York: Springer.

Levi, Isaac (1980). *The Enterprise of Knowledge*. Boston: The MIT Press.

Lewis, C.I. (1946). *An Analysis of Knowledge and Valuation*. La Salle, IL: Open Court.

Lewis, David (1976). Probabilities of Conditionals and Conditional Probabilities. *The Philosophical Review* 85, pp. 297–315.

Lewis, David (1980). A Subjectivist's Guide to Objective Chance. In: *Studies in Inductive Logic and Probability*. Ed. by Richard C. Jeffrey. Vol. 2. Berkeley: University of California Press, pp. 263–94.

Lewis, David (1994). Humean Supervenience Debugged. *Mind* 103, pp. 473–90.

Lindley, Dennis V. (1985). *Making Decisions*. 2nd edition. London: Wiley.

Locke, John (1689/1975). *An Essay Concerning Human Understanding*. Ed. by Peter H. Nidditch. Oxford: Oxford University Press.

Makinson, David C. (1965). The Paradox of the Preface. *Analysis* 25, pp. 205–7.

Makinson, David C. (2011). Conditional Probability in the Light of Qualitative Belief Change. *Journal of Philosophical Logic* 40, pp. 121–53.

Mazurkiewicz, Stefan (1932). Zur Axiomatik der Wahrscheinlichkeitsrechnung. *Comptes rendues des séances de la Société des Sciences et des Lettres de Varsovie* 25, pp. 1–4.

Meacham, Christopher J. G. (2010b). Two Mistakes Regarding the Principal Principle. *British Journal for the Philosophy of Science* 61, pp. 407–31.

Meacham, Christopher J. G. (2016). Ur-Priors, Conditionalization, and Ur-Prior Conditionalization. *Ergo* 3, pp. 444–92.

Moss, Sarah (2018). *Probabilistic Knowledge*. Oxford: Oxford University Press.

Papineau, David (2012). *Philosophical Devices: Proofs, Probabilities, Possibilities, and Sets*. Oxford: Oxford University Press.

Pascal, Blaise (1670/1910). *Pensées*. Translated by W.F. Trotter. London: Dent.

Pearson, K., A. Lee, and L. Bramley-Moore (1899). Genetic (Reproductive) Selection: Inheritance of Fertility in Man. *Philosophical Transactions of the Royal Society A* 73, pp. 534–39.

Peirce, Charles Sanders (1910/1932). Notes on the Doctrine of Chances. In: *Collected Papers of Charles Sanders Peirce*. Ed. by Charles Hartshorne and Paul Weiss. Cambridge, MA: Harvard University Press, pp. 404–14.

Pettigrew, Richard and Michael G. Titelbaum (2014). Deference Done Right. *Philosophers' Imprint* 14, pp. 1–19.

Pollock, John L. (2001). Defeasible Reasoning with Variable Degrees of Justification. *Artificial Intelligence* 133, pp. 233–82.

Popper, Karl R. (1938). A Set of Independent Axioms for Probability. *Mind* 47, pp. 275–9.

Popper, Karl R. (1955). Two Autonomous Axiom Systems for the Calculus of Probabilities. *British Journal for the Philosophy of Science* 6, pp. 51–7.

Popper, Karl R. (1957). The Propensity Interpretation of the Calculus of Probability and the Quantum Theory. *The Colston Papers* 9. Ed. by S. Körner, pp. 65–70.

Ramsey, Frank P. (1929/1990). General Propositions and Causality. In: *Philosophical Papers*. Ed. by D.H. Mellor. Cambridge: Cambridge University Press, pp. 145–63.

Ramsey, Frank P. (1931). Truth and Probability. In: *The Foundations of Mathematics and other Logic Essays*. Ed. by R.B. Braithwaite. New York: Harcourt, Brace and Company, pp. 156–98.

Reichenbach, Hans (1935/1949). *The Theory of Probability*. English expanded version of the German original. Berkeley: University of California Press.

Reichenbach, Hans (1938). *Experience and Prediction*. Chicago: University of Chicago Press.

Reichenbach, Hans (1956). The Principle of Common Cause. In: *The Direction of Time*. Berkeley: University of California Press, pp. 157–60.

Renyi, Alfred (1970). *Foundations of Probability*. San Francisco: Holden-Day.

Roeper, P. and H. Leblanc (1999). *Probability Theory and Probability Logic*. Toronto: University of Toronto Press.

Salmon, Wesley (1966). *The Foundations of Scientific Inference*. Pittsburgh: University of Pittsburgh Press.

Schwarz, Wolfgang (2018). Subjunctive Conditional Probability. *Journal of Philosophical Logic* 47, pp. 47–66.

Seidenfeld, Teddy (1986). Entropy and Uncertainty. *Philosophy of Science* 53, pp. 467–91.

Seidenfeld, Teddy, M.J. Schervish, and J.B. Kadane (2017). Non-Conglomerability for Countably Additive Measures That Are Not κ-Additive. *The Review of Symbolic Logic* 10, pp. 284–300.

Selvin, Steve (1975). A Problem in Probability. *The American Statistician* 29. Published among the Letters to the Editor, p. 67.

Shapiro, Amram, Louise Firth Campbell, and Rosalind Wright (2014). *The Book of Odds*. New York: Harper Collins.

Simpson, E.H. (1951). The Interpretation of Interaction in Contingency Tables. *Journal of the Royal Statistical Society, Series B* 13, pp. 238–41.

Skyrms, Brian (1980b). Higher Order Degrees of Belief. In: *Prospects for Pragmatism*. Ed. by D.H. Mellor. Cambridge: Cambridge University Press, pp. 109–37.

Skyrms, Brian (2000). *Choice & Chance: An Introduction to Inductive Logic*. 4th edition. Stamford, CT: Wadsworth.

Spohn, Wolfgang (2012). *The Laws of Belief: Ranking Theory and its Philosophical Applications*. Oxford: Oxford University Press.

Stephenson, Todd A. (2000). *An Introduction to Bayesian Network Theory and Usage*. Tech. rep. 03. IDIAP.

Talbott, William J. (2016). Bayesian Epistemology. In: *The Stanford Encyclopedia of Philosophy*. Ed. by Edward N. Zalta. Winter 2016. Metaphysics Research Lab, Stanford University. URL: https://plato.stanford.edu/archives/win2016/entries/epistemology-bayesian/.

Teller, Paul (1973). Conditionalization and Observation. *Synthese* 26, pp. 218–58.

Titelbaum, Michael G. (2013a). *Quitting Certainties: A Bayesian Framework Modeling Degrees of Belief*. Oxford: Oxford University Press.

Tversky, Amos and Daniel Kahneman (1974). Judgment under Uncertainty: Heuristics and Biases. *Science* 185, pp. 1124–31.

Tversky, Amos and Daniel Kahneman (1983). Extensional Versus Intuitive Reasoning: The Conjunction Fallacy in Probability Judgment. *Psychological Review* 90, pp. 293–315.

van Fraassen, Bas C. (1981). A Problem for Relative Information Minimizers. *British Journal for the Philosophy of Science* 32, pp. 375–79.

van Fraassen, Bas C. (1982). Rational Belief and the Common Cause Principle. In: *What? Where? When? Why?* Ed. by Robert McLaughlin. Dordrecht: Reidel, pp. 193–209.

van Fraassen, Bas C. (1984). Belief and the Will. *The Journal of Philosophy* 81, pp. 235–56.

van Fraassen, Bas C. (1989). *Laws and Symmetry*. Oxford: Clarendon Press.

van Fraassen, Bas C. (1995). Belief and the Problem of Ulysses and the Sirens. *Philosophical Studies* 77, pp. 7–37.

van Fraassen, Bas C. (1999). Conditionalization: A New Argument For. *Topoi* 18, pp. 93–6.

Venn, John (1866). *The Logic of Chance*. London, Cambridge: Macmillan.

von Mises, Richard (1928/1957). *Probability, Statistics and Truth*. (English edition of the original German *Wahrscheinlichkeit, Statistik und Wahrheit*.) New York: Dover.

Wainer, Howard (2011). *Uneducated Guesses: Using Evidence to Uncover Misguided Education Policies.* Princeton, NJ: Princeton University Press.

Weatherson, Brian and Andy Egan (2011). Epistemic Modals and Epistemic Modality. In: *Epistemic Modality.* Ed. by Andy Egan and Brian Weatherson. Oxford: Oxford University Press, pp. 1–18.

Weintraub, Ruth (2001). The Lottery: A Paradox Regained and Resolved. *Synthese* 129, pp. 439–49.

Weisberg, Jonathan (2007). Conditionalization, Reflection, and Self-Knowledge. *Philosophical Studies* 135, pp. 179–97.

Weisberg, Jonathan (2009). Varieties of Bayesianism. In: *Handbook of the History of Logic.* Ed. by Dov. M Gabbya, Stephan Hartmann, and John Woods. Vol. 10: Inductive Logic. Oxford: Elsevier.

White, Roger (2005). Epistemic Permissiveness. *Philosophical Perspectives* 19, pp. 445–59.

Williams, J. Robert G. (2016). Probability and Non-Classical Logic. In: *Oxford Handbook of Probability and Philosophy.* Ed. by Alan Hájek and Christopher R. Hitchcock. Oxford: Oxford University Press.

Williamson, Timothy (2007). How Probable Is an Infinite Sequence of Heads? *Analysis* 67, pp. 173–80.

Wittgenstein, Ludwig (1921/1961). *Tractatus Logico-Philosophicus.* Translated by D.F. Pears and B.F. McGuinness. London: Routledge.

Yalcin, Seth (2012). A Counterexample to Modus Tollens. *Journal of Philosophical Logic* 41, pp. 1001–24.

Yule, G.U. (1903). Notes on the Theory of Association of Attributes in Statistics. *Biometrika* 2, pp. 121–34.

Index of Names in Volume 1